Molecular A

Eman Jarallah
Fatima Moeen Abbas

Molecular Analysis of Carbapenem Resistance

in clinical isolates of Klebsiella pneumoniae

Scholars' Press

Publisher:
Scholars' Press
is a trademark of
International Book Market Service Ltd., member of OmniScriptum Publishing Group
17 Meldrum Street, Beau Bassin 71504, Mauritius
Printed at: see last page
ISBN: 978-3-659-83852-1

Molecular Analysis of Carbapenem Resistanc Clinical *Klebsiella pneumoniae*

By

Dr.Eman Mohammad Jarallah

and

Dr.Fatima Moeen Abbas

Summary

This study analyzed the prevalence of carbapenem resistance in clinical isolates of *Klebsiella pneumoniae*.

During the period from April to August 2011, a total of 801 various clinical samples were collected from different hospitals in Hilla city. *K.pneumoniae* isolates were identified by standard biochemical tests ,then confirmed by VITEK 2 system . Antibiotic susceptibility profile and minimum inhibitory concentration (MIC) tests were determined by the disk diffusion and HiComb, Minimum inhibitory concentration evaluator (M.I.C.E.) methods, respectively. The ability of isolates to produce β - lactamase was determined by iodometric method. Carbapenemase, extended –spectrum β - lactamase (ESBL) and AmpC β - lactamase were initially detected and confirmed by the phenotypic methods. Carbapenemase –encoding genes (KPC,IMP,VIM,NDM-1, OXA-23 and SME), ESBLs genes (TEM, SHV, CTX-M, OXA-1,VEB,PER and GES) and AmpC β - lactamases (AmpC) were detected by conventional Polymerase chain reaction (PCR) technique. Conjugation experiments were performed by using rifampicin resistant *Escherchia coli* MM294 as a recipient strain.

Out of 801 ,177 *Klebsiella* spp. isolates were identified .Of these,117 were specified as *K.pneumoniae* . High prevalence were detected in Babylon Teaching Hospital for Maternity and Pediatric 65 (23%) ,followed by Chest Diseases Center 19 (14.8%). All *K.pneumoniae* isolates were identified to the level of subspecies. Results revealed that 77 (9.6%),34 (4%) and 6 (1%) were belonged to *K.pneumoniae* subsp. *pneumoniae* , *K.pneumoniae* subsp. *ozaenae* and *K.pneumoniae* subsp *rhinoscleromatis*, respectively.

All 117 *K.pneumoniae* isolates were primarily screened for β-lactams resistance, 91 (78%) were found to be screen positive. All β-lactams resistant isolates were resistant to a minimum of three classes of antibiotics, hence these isolates were considered as multidrug resistance. Out of 91 β-lactams resistant isolates, 77(84.6%) were detected as β-lactamase producers by rapid iodometric method.

All 91 β-lactams resistant isolates were found to be ESBL producers, initially. In confirmatory tests, CHROMagar technique was the most efficient method than disk approximation method.

Out of the 91 β-lactam resistant isolates, 17 (18.7%) were identified as resistant to carbapenem antibiotics by disk diffusion method, 9 (10%) for imipenem,16(17.6%) for meropenem and 17 (18.7%) for ertapenem. All isolates were subjected to three confirmatory tests for carbapenemase production. Phenotypic detection of carbapenemases by imipenem-EDTA disk, modified Hodge test and KPC CHROMagar identified a proportion of (65%), (82%) and (100%) as carbapenemase producers ,respectively. PCR analysis confirmed 3 different carbapenemase genes bla_{NDM-1}, bla_{VIM} and bla_{OXA-23} in percentage of 17.6% ,82.3%,and 88.2%, respectively.

All carbapenemase positive isolates were tested for antibiotic susceptibility towards 26 antibiotics and for MIC against selected β-lactam antibiotics .Results showed that 11(65%) and 6(35%) of the isolates were Extensive drug resistant (XDR) and Pandrug-resistant (PDR), respectively. Most carbapenemase positive isolates were resistant to imipenem, meropenem, ampicillin cefotaxime,ceftriaxone and ceftazidime with concentrations reached beyond the break point values .

The PCR results confirmed that 13 (76.5%) of carbapenemase positive isolates harbored bla_{TEM}, bla_{SHV}, bla_{CTX-M}, bla_{OXA-1} and only 10

(58.8%) isolates harbored *bla*PER genes. While none of the isolates showed amplification of either *bla*VEB or *bla*GES genes.

All carbapenemase positive isolates were further tested for potential AmpC production by cefoxitin susceptibility and confirmed by two phenotypic (modified –three dimensional and AmpC disk) tests. Results indicated that all carbapenemase positive isolates were cefoxitin resistant and 3 (17.6%), 2(11.8%) were AmpC producers by two above tests, respectively . Whereas none of the isolates were positive for inducible AmpC β - lactamase by ceftaizdime-imipenem antagonism test. *Bla*AmpC was detected in 13 (76.5%) of the isolates.

The carbapenemase positive *K.pneumoniae* isolates K13 and K14 (carried *bla*VIM, *bla*NDM-1 and *bla*OXA-23) were investigated for their ability to transfer carbapenem resistance. Conjugation experiments and PCR assay demonstrated that the two isolates were able to transfer *bla*VIM in transconjugants. Antibiotic susceptibilities of the two transconjugants showed that the MICs of imipenem and meropenem was much higher (>32 μg /ml) relative to those of the recipient . The PCR identified that the two isolates were able to transfer *bla*SHV, *bla*CTX-M and *bla*AmpC genes in transconjugants.

These results point out that carbapenemase producing *K.pneumoniae* were detected in both phenotypic and molecular methods (PCR). This underlies the importance of their accurate identifications and reporting to prevent the emergence of complete resistance to the most potent drugs against *K.pneumoniae* in Hilla.

List of Contents

Chapter Three: Materials and Methods   37

List of Tables

List of Figures

XI

List of Abbreviations

General Abbreviations

Abbreviation	Key
AME	Aminoglycoside modifying enzyme
AmpC	Molecular class C β - lactamase
ATCC	American type culture collection
bla gene	β - lactamase gene
CAZ	Ceftazidime
CHROMagar	Chromogenic orientation agar
CIR	Carbapenem Intermediate /Resistant
CIs	Chromosomal integrons
CLSI	Clinical and Laboratory Standards Institue
CRS	Carbapenem Reduced Susceptible
CTX	Cefotaxime
CTX-M	Cefotaximase, β - lactamase active on cefotaxime
Cys	Cystein
D.D.W.	Deionized sterile distilled water
D.W.	Distilled water
DNA	Deoxyribonucleic acid
EDTA	Ethylene diaminetetraacetic acid
ESBL	Extended-spectrum β - lactamase
GES	Guiana extended spectrum β - lactamase
GIM	German imipenemase
HCl	Hydrochloric acid
ICU	Intensive care unit
IMP	Imipenemase, β - lactamase active on imipenem
IMI/NMC	Imipenem- hydrolyzing β - lactamase /not metalloenzyme carbapenemase
IS	Insertion sequence
K	Capsular antigen
KIA	Kligler iron agar
KPC	*Klebsiella pneumoniae*- carbapenemase
LB	Luria – Bertani

LPS	Lipopolysaccharide
MBL	Metallo- β - lactamase
MDR	Multi- drug resistance
MHA	Mueller- Hinton agar
MHT	Modified Hodge test
MIC	Minimum inhibitory concentration
M.I.C.E.	Minimum inhibitory concentration evaluator
MIs	Mobile integrons
MTDT	Modified three- dimensional test
NAG	N-acetylglucosamine
NAM	N-acetylmuramic acid
NCCLS	National Comittee for Clinical Laboratory Standards
NDM	New Delhi metallo β - lactamase
OMP	Outer membrane protein
OXA	Oxacillinases , β - lactamase active on oxacillin
PMABLs	Plasmid- mediated AmpC β - lactamases
PBP	Penicillin binding protein
PBS	Phosphate buffer solution
PCR	Polymerase chain reaction
PDR	Pandrug resistance
PER	Pseudomonas extended resistant
PH	Power of hydrogen (H^+)
SDS	Sodium dodecyl sulfate
SET	Salt- EDTA – tris buffer
SHV	Sulfhydryl variable β - lactamase
SIM	Seoul imipenemase
SIs	Super- integrons
SME	*Serratia marcescens* enzyme
SPM	Sao Paulo metallo- β - lactamase
Spp	Species
TBE	Tris borate – EDTA buffer
TE	Tris – EDTA buffer
TEM	β - lactamase named after first patient isolated from

	Temarian.
Tris-OH	Tris- (Hydroxymethyl) methylamine
VEB	Vietnam extended- spectrum β - lactamase
VIM	Verona integron- encoded metallo- β - lactamase
XDR	Extensive-drug resistance

1.Introduction

Although a ubiquitous pathogen ,yet capable of causing both community and healthcare-associated infections that range from mild urinary tract infection to sever bacteremia and pneumonia, *Klebsiella pneumoniae* has a remarkable ability to up regulate or acquire resistance determinants, making it one of the most feared organisms threatening the current antibiotic era (Brisse *et al* ,2006;Robin *et al*.,2011 and Cabral *et al*.,2012).It has recently emerged as a major cause of hospital- acquired infections because of its propensity to accumulate mechanisms of antimicrobial resistance leading to pandrug resistance , and causing outbreaks that often involve multiple facilities in hospitals (Bush,2010 and Snitkin *et al*.,2012).

Antimicrobial resistance is a growing threat worldwide. The foundation of modern medicine is built on the availability of effective antibiotics, especially in economically deprived areas of the world where the disease burden due to bacterial infections remain high .Antibiotic resistance is predominantly fueled by antibiotic use (Goossens *et al*.,2005) .With their broad spectrum activity for Gram-positives , Gram-negatives and anaerobic bacteria ,carbapenems are frequently used as a last line of therapy (Ejikeuqwu *et al*.,2012 and Kosmidis *et al*.,2012).

Carbapenem antibiotics play a crucial role in the treatment of serious nosocomial infections caused by organisms with reduced susceptibility to other antimicrobials (Patel *et al*.,2011) . Unfortunately ,the prevalence of carbapenem- resistant isolates appears to be increasing .These isolates are difficult to control because they spread easily within and between hospitals (Kochar *et al*.,2009),and treatment options for infections caused by carbapenem - resistance isolates are limited and are associated with mortality rates upwards of 50% (Patel *et al*.,2008 and Ben-David *et al*.,2012). Besides the enzymatic mechanism like carbapenemase

1

production (serine carbapenemase or metallo-β-lactamase), carbapenem resistance has also been ascribed to nonenzymatic mechanisms like hyperproduction of AmpC cephalosprinases ,or extended spectrum β-lactamases (ESBLs), in combination with the loss of major porins, augmented drug efflux and alteration in penicillin-binding proteins (PBPs) (Queenan and Bush,2007;Wang *et al.*,2009 and Patel and Bonomo,2011).

Identification of risk factors for infection with carbapenem - resistant *K.pneumoniae* may help in the empirical therapeutic decision-making process and may assist in the early implementation of appropriate infection control measures (Patel *et al.*,2008). However, increased risk of colonization and/or infection with carbapenem - resistant *K.pneumoniae* has been identified for patients with poor functional states (Schwaber *et al.*, 2008), sever illness (Gasink *et al.*,2009) and prolonged hospitalization (Wiener-Well *et al.*,2010), patients admitted to intensive care units (ICUs) (Falagas *et al.*,2007) and patients exposed to healthcare-associated risk factors such as organ or stem-cell transplantation (Patel *et al.*,2008), mechanical ventilation, surgery, transfer between units (Falagas *et al.*,2007 and Gregory *et al.*,2010) and antecedent treatment with different antibiotics (Kwak *et al.*,2005 and Hussein *et al* .,2009).

The emergence of carbapenem –resistant *K.pneumoniae* is a great public health concern because there is no reliable treatment. They are typically resistant not only to carbapenem,but also to polymyixin B sulfate and third generation cephalosporins .In addition, many carbapenem –resistant isolates of *K.pneumoniae* also possess ESBLs, and genes conferring resistance may be accompanied by virulence factors (Bratu *et al* .,2005b and Schwaber and Carmeli,2008).

However,in Hilla city there are no data on the prevalence of carbapenem resistance in clinical isolates of *K.pneumoniae* .Hence, the present study aims to determine the prevalence of carbapenemases and other β -lactamase genes responsible for carbapenem –resistance among clinical isolates of *K.pneumoniae* collected from different hospitals in Hilla city. For this aim the following steps were achieved :

1-Isolation and identification of *K.pneumoniae* from different clinical samples in Hilla hospitals .

2-Determining the prevalence of carbapenem resistant clinical isolates of *K.pneumoniae.*

3-Detecting the genes responsible for carbapenemase production of KPC, VIM, IMP, NDM-1, OXA-23 and SME among carbapenem-resistant isolates.

4-Investigating the dissemination of ESBL genes (*bla*TEM, *bla*SHV, *bla*CTX-M, *bla*OXA-1, *bla*PER, *bla*VEB ,*bla*GES) and AmpC β - lactamases in carbapenemase producing isolates.

5-Evaluating the mobility of carbapenemase, ESBL and AmpC β - lactamase genes by transconjugation experiment.

2. Literature Review

2.1.Genus *Klebsiella*

Genus *Klebsiella* is among the oldest known genera in the family *Enterobacteriaceae*, described for the first time by Trevisan in 1885 in honor of the German microbiologist Edwin Klebs (1834-1913), the first *Klebsiella* strain ever described was a capsulated bacillus from a patient with rhinoscleroma (Brisse *et al*., 2006).

The genus *Klebsiella* is Gram – negative , non – motile, non-sporulating , lactose- fermenting , oxidase negative with a prominent polysaccharide capsule. *Klebsiella* are every where in nature , while in human they colonize the skin, pharynax, or gastrointestinal tract, sterile wound, urine and may be regarded as normal flora in many parts of the colon, intestinal and biliary tract (Podschun and Ullman, 1998 and Brisse *et al*., 2006).

Klebsiella spp. are opportunistic human pathogens that can be isolated from various animal and human clinical specimens, and responsible for a wide range of clinical syndromes including purulent infection, urinary tract infection, pneumonia, septicemia and meningitis (Younes *et al*., 2011). They have been prominent among Gram- negative bacilli causing nosocomial infections, as well as being an important source of transferable antibiotic resistance (Miranda *et al*., 2004).

2.1.1.Taxonomy

K.pneumoniae is a member of *Enterobacteriaceae* family, recognized over 100 years ago as a cause of community- acquired pneumonia (Keynan and Rubinstein , 2007). Originally , the medical importance of the genus *Klebsiella* led to its being subdivided into three species corresponding to the diseases they caused: *K. pneumoniae*, *K. ozaenae*, and *K. rhinoscleromatis*. Based on DNA- DNA hybridization

4

data, *K. ozaenae* and *K. rhinoscleromatis* , taxonomically , are regarded as subspecies of *K. pneumoniae* (Podsschun and Ullmann, 1998; Drancourt *et al.*, 2001 and Hansen *et al.*, 2004).

K. oxytoca was further considered as a distinct group from *K. pneumoniae* (Jain *et al.*, 1974). Other four species : *K. planticola* , *K. terrigena* , *K.trevisanii*, and *K. ornithinolytica* were identified in 1980s by (Bagley *et al.*, 1981 ; Izard *et al.*, 1981 ; Ferragut *et al.*, 1983 and Sakazaki *et al.*, 1989) respectively .The *K. planticola* and *K. trevisanii* were subsequently combined and considered as *K. planticola* on the basis of DNA – DNA hybridization (Gavini *et al.*,1986). *Enterobacter aerogenes* and *Calymmatobacterium granulomatis*, due to their close relationships to *Klebsiella* species were regarded as an eighth and ninth member of the genus *Klebsiella* and named *K. mobilis* and *K. granulomatis* respectively (Carter *et al.*, 1999 and Drancourt *et al.*, 2001). The three species :*K. terrigena, K. planticola* , and *K. ornithinolytica* have recently been transferred to the new genus *Raoultella* (Drancourt *et al.*, 2001).

Clinical isolates of *K.pneumoniae* fall into four phylogentic groups based on nucleotide variations of the *gyrA* , *parC*, and *rpoB* genes named KpI , KpII-A , KpII-B and KpIII , with the newly described species *K. variicola* appearing to correspond to KpIII (Alves *et al.*,2006).

2.2.*Klebsiella pneumoniae*

2.2.1.General Characteristics

K.pneumoniae is rod-shaped bacterium 0.3-1 μm in diameter and 0.6-6 μm in length, arranged singly, in pairs or in short chains, the (G + C) content in DNA molecule range from (53-58) mole percent. Gram-negative, non- motile with a prominent polysaccharides capsule of considerable thickness which gives the colonies their glistening and mucoid appearance on agar plates (Holt *et al.*, 1994). It is facultative anaerobic, chemoorganotrophic bacteria,having both a respiration and a fermentative type of metabolism. Optimal temperature for growth is 37 C^o and are killed by moist heat at 55 C^o for 30 min, may survive drying for months and, when kept at room temperature, cultures remain viable for many weeks. On MacConkey's agar, the colonies appear large ,mucoid, and red with diffusing red pigment indicating fermentation of lactose and acid reduction (Koneman *et al.*,1994;Holt *et al.*, 1994 and Chart, 2007).

K. pneumoniae produce lysine decarboxylase but not ornithine decarboxylase and arginine dihydrolase. In addition, it hydrolyzes urea slowly, producing a light pink color on the slant of Christensen's urea agar .Production of indole from tryptophan is negative. It reduces nitrates ,gives oxidase negative and catalase positive. It usually gives positive test for citrate and Voges-Proskauer reactions (Brooks *et al.*,2001).

2.2.2.Pathogenesis and Clinical Importance

K. pnenmoniae is the most medically important species of the genus *Klebsiella*. In recent years , *Klebsiella* have become important pathogens in nosocomial infections (Alves *et al.*, 2006). It is also a potential community – acquired pathogen (Ko *et al* ., 2002).

Klebsiella is an important opportunistic pathogen, can cause infections of respiratory tract , nasal mucosa , pharynx and generally results in primary pneumonia, septicemia and urinary tract infection (Sikarwar and Batra , 2011). Pneumonia is the most frequent nosocomial infection (30 to 33% of cases) among combined medical- surgical intensive care units participating in the National Nosocomial Infections Surveillance System (Richards *et al* ., 2000).

Klebsiella species were found to be the most frequently isolated Gram negative bacteria in cases of primary bacteremia (Cross *et al.*, 1983). It is the second pathogen, next to *Escherchia coli* that causes urinary tract infection . Several other reports have also observed that this pathogen had become the predominant cause of liver abscess instead of the previously described *E. coli* , Streptococci, anaerobic bacteria and amoebae (Yeoh *et al.*, 1997 ; Ohmori *et al.*, 2002 and Lederman and Crum, 2005). *Klebsiella* species especially *K.pneumoniae* has been shown to cause intra-abdominal infections mediated by heat stable and heat labile enterotoxins (Rennie *et al.*,1990).Moreover, *K.pneumoniae* was associated with chronic diarrhoea in HIV-infected persons (Nguyen *et al.*, 2003).

K. pneumoniae has also become a common cause of community acquired bacterial meningitis in adults in Taiwan. The proportion of cases of bacterial meningitis due to *K. pneumoniae* in one Taiwanese hospital increased from 8% during 1981 through 1986 to 18% during 1987 through 1995 (Tang, *et al.* 1997). Outside Taiwan, cases of *K. pneumoniae* meningitis have occurred, predominantly in other parts of Asia (Yanagawa, *et al.* 1989; Li *et al.*, 2001 and Ohmori, *et al.* 2002), Europe and North America (Holder and Halkias 1988; Giobbia, *et al.* 2003; Bouadma, *et al.* 2006 and Braiteh and Golden ,2007). The rarity of these cases outside Asia raises the possibility of ethnicity or country of origin predisposing individuals to invasive disease (Ko *et al.*,2002).

As an opportunistic pathogen, it normally affects persons with low immune system such as hospital patients , diabetes patients and people with chronic lung disease . Many times , alcoholics also suffer from *K.*

pneumoniae infections. Thus, the infections are either hospital- acquired or community- acquired (Sikarwar and Batra , 2011).

Numerous reports have been published worldwide on outbreaks caused by *K. pneumoniae* in different healthcare settings, like neonatal wards , nursing homes, and intensive care units (Lytsy *et al.*, 2008 and Cai *et al.*,2011).

2.2.3.Epidemiology

K.pneumoniae is an opportunistic and major hospital –acquired pathogen that has the potential to cause sever morbidity and mortality, particularly in intensive care units and amongst pediatrics patients ,but also in medical and surgical wards (Branger *et al.*,1998;Decre *et al.*,1998 and Podschun and Ullmann, 1998). *Klebsiella* is the cause of 5-7.5 % of all nosocomial infections and the third most –common bacterial cause of hospital -acquired pneumonia (Jones,2010 and Khorshidi *et al.*,2012).

However, the intestine is one of the major reservoirs of *K. pneumoniae*, and epidemiological studies have suggested that the majority of *K. pneumoniae* infections are preceded by colonization of the gastrointestinal tract (Lin *et al* .,2012). Frequency of reservoirs of these bacteria increases dramatically in the hospital where colonization has direct relation with length of hospital stay (Khorshidi *et al* .,2011). According to Selden *et al* (1971) investigations ,the rate of reservoirs of *Klebsiella* among hospitalized patients is nearly 77 % with colonization in feces ,19 % in pharynax,42 % on hands which is directly associated with antibiotics administration.

K.pneumoniae is the most common strain found in hospitalized patients and has been reported to cause outbreak of sepsis and death of newborns in the intensive care unit of a tertiary hospital like in the United Kingdom (Johnson *et al* .,1992), in Brazil (Otman *et al.*,2002) ,and in the

Middle East (Tamma *et al.*,2012). The observed mortality rates range from about 25-50 % (Feldman *et al* .,1995) .Mortality rates are as high as 50 % and approach 100 % in hospitalized immunocompromised patient with underlying diseases such as diabetes mellitus (Ko *et al.*,2002). However, in a local study in Hilla, *K.pneumoniae* was the most dominant species in pediatric bloodstream infections (Al-Asady,2009). It was predominant isolates in clinical and local hospital environmental samples (Al-Charrakh ,2005 and Al-Hilli,2010).

2.3. β - Lactam Antibiotics

The β lactam antibiotics are the largest and most commonly used group of antimicrobial agents for treatment of bacterial infections. They are a broad class of antibiotics, consisting of all antibiotic agents that contains a β – lactam nucleus in its molecular structure, hence the name (Beg *et al.*, 2011).

The β - lactam antibiotics are bactericidal cell wall synthesis inhibitors , owing to their high effectiveness , broad spectra and low toxicity, low cost , ease of delivery and minimal side effects (Wilke *et al.*, 2005).

2.3.1.General Structure and Function

β - lactam is a generic name for all β - lactam antibiotics that contain a β -lactam ring , a heteroatomic ring structure, consisting of three carbon atoms and one nitrogen atom (Wilke *et al.*, 2005). Based on their chemical structure they can be divided into six different groups, the pencicllins , cephalosporins, cephamycins, monobactams, carbapenem and β - lactamase inhibitors (Smet *et al.*, 2008).

The basic structure of the penicillins is a thiazolidine ring connected to a β - lactam ring, to which an acyl side chain is attached. The side chain of different penicillins is modified in order to change the activity (Soares *et al.*,2012). The penicillins are all sensitive to the β -lactamases, however, temocillins , is resistant to most of these enzymes (Livermore *et al.*, 2006).

Cephalosporin compound were first isolated from cultures of *Cephalosporium acremonium* in 1948 by the Italian scientist Giuseppe Brotzu. He noticed that these cultures produced substances that were effective against *Salmonella typhi*, the cause of typhoid fever. The cephalosporin nucleus, 7- aminocephalosporanic acid (7-ACA), was derived from cephalosporin C and proved to be analogous to the penicillin nucleus 6- amino penicillanic acid. Modification of the 7- ACA side- chains resulted in the development of useful antibiotic agent, and the first agent cephalothin (cefalotin) was launched by Eli Lilly in 1964. Cephalosporins are bactericidal and have the same mode of action as other β - lactam antibiotics (such as penicillins). Cephalosporins distrup the synthesis of peptidoglycan layer of bacterial cell walls (Beg *et al.*,2011*)*.

In monobactams the β - lactam ring is alone, and not fused to another ring , hence they lack the double ring structure found in traditional β - lactam antibiotics and they can be easily synthesized. The monobactams like aztreonam have different side chains affixed to monocyclic nucleus. Monobactams have a wide range of activity to aerobic Gram- negative bacteria (Bush, 1996).

Carbapenems are another class of β - lactam antibiotics which are produced by *Streptomyces cattylea* (Medeiros, 1997). The term carbapenem is defined as the 4:5 fused ring lactam of penicillins with a double bond between C-2 and C-3, but the sulplur atom at position -1 of the structure has been replaced with a carbon atom and hence the name of the group, the carbapenems (Papp- Wallace *et al.*, 2011). They came into use in the 1980s; and consist of imipenem, meropenem, doripenem, ertapenem, panipenem and biapenem with an exceptionally broadest spectrum of activity and greatest potency against Gram- positive , Gram-negative and anaerobic organisms (Prakash, 2006 and Papp- Wallace *et al.*,2011).

The hydroxyethyl side chain in it's transconfiguration at position 6 confers stability toward most β - lactamases, including the extended spectrum β - lactamases (ESBLs) and AmpC β - lactamase produced by Gram- negative bacilli (Bonfiglio *et al.*, 2002 and Prakash, 2006).As a result, they are often used as " last – line agents " "or antibiotics of last resort" for the treatment of severe infections caused by multidrug-resistant AmpC- or extended- spectrum β - lactamase- producing *Enterobacteriaceae* isolates (Queenan and Bush, 2007 and Tailor,2011).

Carbapenems have a high degree of toxicity and their mode of action is by inhibition of cell wall synthesis in bacteria by binding to penicillin – binding proteins (PBPs) (Dugal and Fernandes, 2011).PBPs are enzymes (transglycolases, transpeptidases and carboxypeptidases) that catalyze the formation of the pepdidoglycan in the cell wall of bacteria (Papp-Wallace *et al.*, 2011).A key factor of the efficacy of carbapenems is their ability to bind to multiple different PBPs (Hashizume *et al.*,1984). Imipenem binds preferentially to PBP2, followed by PBP1 a and 1 b. Meropenem and ertapenem bind most strongly to PBP2, followed by PBP3 then PBP1 a and 1b. Doripenem has been reported to have strong affinity for species – specific PBP targets; PBP2 in *E. coli* (Tailor,2011).

11

β - lactamase inhibitors are designed to inhibit or destroy the effectiveness of β - lactamase enzymes(Forssten,2009). The first inhibitor of β - lactamase which was available for clinical use in 1984 was clavulanic acid. It is packaged in combination with ampicillin and is a natural product of *Streptomyces clavuligerus*. The second group of inhibitors the penicillianc acid sulfones, sulbactam and tazobactam are semisynthetic derivatives of penicillanic acid. These inhibitors bind, acylate the enzyme, and inactivate the active-site of serine in the class A β - lactamases, thus preventing the β - lactamases from hydrolyzing the penicillins in the drug combination (Medeiros, 1997).

β - lactamase inhibitors have variable inhibitory activity against ESBL enzymes (Pfeifer *et al.*, 2010). Tazobactam which appears to be the most potent of the 3 , is active against some of the TEM, SHV and CTX-M enzymes (Payne *et al.*, 1994).

2.3.2.Mode of Action

β - lactam antibiotics are bactericidal agents which inhibit cell wall synthesis (Bradford, 2001). They act on bacteria by binding to so called PBPs which are enzymes involved in cell wall synthesis (Sauvage *et al* ., 2008). The PBPs are located on the outer side of the cytoplasmic membrane. In Gram – negative bacteria this is in the periplasmic space. These antibiotics interfere with the structural cross linking of peptidoglycans and as such preventing terminal transpeptidation in the bacterial cell wall (Andes and Craig , 2005).

The bacterial cell wall is a complex structure formed by cross- links of peptidoglycan, maintaining cell shape and structure (Tailor,2011). The peptidoglycan is composed of a basic repeating unit of an alternating chains of disaccharide *N*-acetylglucosamine (NAG) and *N*-acetylmuramic acid (NAM) linked by β-(1,4)-glycoside units. The carboxyl group of muramic acid is usually replaced by an amino acid chain composed of four amino acids. The most common are L-alanine, D-alanine, D-glutamic acid, D-glutamine and L-lysine or 5 diaminopimelic acid (DAP) (Tipper and Strominger 1965 and Livermore and Williams1996).

The mechanism of action is explained by the structural similarity between the beta-lactam ring and the peptidoglycan bluiding block acyl-D- alanyl- D- alanine (Tipper and Strominger, 1965). In the presence of the β-lactam antibiotics, the transpeptidases and carboxypeptidases react with acyl-D-alanyl-D-alanine to form alethal serine-ester-linked acyl (penicilloyl, cephalosporoyl) enzyme complex. The β-lactam-enzyme complex is very stable, and blocks the normal transpeptidation reaction. This result disrupts the synthesis of the cell wall and makes the growing bacteria highly susceptible to cell lysis and death (Tipper and Strominger 1965; Ghuysen 1988; Livermore and Williams 1996 and Wilke, *et al.* 2005).

2.3.3.Mechanisms of Resistance to β - Lactam Antibiotics

Bacterial resistance to antibiotics may be either natural (intrinsic, which refers to an inherent resistance to an antibiotic that is a naturally occurring feature of the microorganism), or acquired (this is usually due to mutation).

Acquired resistance usually takes place suddenly in microorganisms. It involves chromosomal mutation which may be due to microbial ability to acquire certain transferable gene such as plasmid, transposon, and integrons which confer on them resistance to antimicrobial agents. Once a microorganism had acquired resistance to a particular drug , it is possible for such ability to be transferred to other organisms vi either one of horizontal gene transfer system or vertical gene transfer method (Elufisan *et al.*, 2012 and Soares *et al.*, 2012).

a) Changes in Target Site

Most antibiotics exhibit their activities on microorganisms by acting on a particular site on the microbes referred to as the drug target site on the microorganisms. Antibiotics usually have a high affinity for the sites on the microbes and can out compete any other substrate that may also depend on this site. However, if this site is modified or altered, it may lead to reduced or loss of affinity for the antibiotics and the bacteria with such altered site becomes resistant to antibiotics.

The altered penicillin binding protein in the cell wall of a particular strain of *Staphylococcus aureus* confer on it resistance to β - lactam antibiotics. A number of target sites can be affected by mutation, such as penicillin- binding proteins. These sites are known as the receptor site of the microorganism and are usually protein molecule (Elufisan *et a l.*, 2012).

b) Reduction in Cellular Permeability

There is a need for antibiotics to overcome certain mechanical barrier presented by the host cell membrane before they can elicit their activities. Mutants in bacteria have been identified to produce protective coat in the bacteria which reduce or inhibit the cellular uptake of antibiotics such as aminoglycosides and β–lactams, chloramphemicol, bacitracin, and Isoniazid . (Elufisan *et al.*, 2012).The outer membrane alterations in *K.pneumoniae* are not decisive factors in increasing resistance to antimicrobial agents, but porin loss cooperates with β-lactamase production to increase resistance to β-lactams (Hernandez-Alles *et al.*,2000).

c) Increased Efflux Pumps

Efflux mechanism involves the removal of materials (wast products or toxins) from the microbial cell and provide a very efficient means for antibiotic resistance (Van Bembeke *et al.*,2000). Bacteria with this mechanism have efflux pumps through which they selectively force out specific antibiotics. Most of them are multidrug transporters that are capable to pump a wide range of unrelated antibiotics- macrolides, tetracyclines , fluoroquinolones- and thus significantly contribute to multidrug resistance (Dzidic *et al*, 2008 and Elufisan *et al.*, 2012).

Synergy between the outer membrane permeability and the active efflux family plays a significant role on the ability of efflux pump to confer resistance (Elnfisan *et al.*, 2012).

d) Enzymatic Inactivation

Enzymatic inactivation of the antibiotic is a common mechanism of resistance often seen with the β- lactamases inactivating penicillins and cephalosporins, acetyltransferases altering chloramphenicol and aminoglycosides and to a minor extent enzymes that metabolize tetracyclines and the macrolides (Pallasch, 2003). The β- lactam antibiotics are inactive once their β- lactam ring is opened , and beta-lactamases hydrolyze these agents to form a linear molecule incapable of binding to their receptors , PBPs (Koch, 2000).

There are more than 300 β- lactamases both chromosomally and plasmid mediated- some of which are of the β- lactam antibiotics except carbapenem and cephamycins (Bardford , 2001).

2.4. β - Lactamases

β - lactamases are the most common cause of bacterial resistance to the medically important β -lactam antibiotics in Gram – negative bacteria (Babic *et al.*, 2006). These enzymes may have a broad – host range and varying hydrolyzing activity. Resistance is usually mediated by β - lactamase enzymes that hydrolysis the β - lactam ring prior to bind to PBPs on bacterial cytoplasmic membrane (Thirapamnethee , 2012).

β - lactamase in Gram– positive bacteria are secreted to the outside membrane environment as exoenzymes. In Gram– negative bacteria they remain in the periplasmic space, where they attack the antibiotic before it can reach its receptor site (Stratton, 2000).

2.4.1. Evolution and Classification

β - lactamases are large, heterogeneous group enzymes, some named for the source organism or place of discovery , while others are named for their preferred substrate (Thomas,2007).

The first plasmid – mediated β - lactamase in Gram- negative , TEM-1, was described in *E.coli* (Datta and Kontomichalou, 1965). Another plasmid- mediated β - lactamases ,known as " SHV-1" (Sulfhydryl variable), was found in *K. pneumoniae* and *E. coli* (Paterson and Bonomo, 2005).

Different classification schemes for these enzymes have been presented since the beginning of the 1970s, based on phenotype, gene or amino acid protein sequences and function (Hall and Barlow , 2005). One of the most used classification schemes is Ambler's based upon amino acid sequences (Ambler *et al.*, 1991). He classified the β - lactamases into four molecular classes, A, B, C and D. Molecular class A, C and D include enzymes that hydrolyze their substrates by forming an acyl

(Penicilloic or cephalosporoic) enzyme through an active site, the three classes share similarity on the protein structure level, which proves that they are derived from a common ancestor (Hall and Barlow , 2004). Whereas class B β - lactamases are metalloenzymes that utilize at least on active- site zinc ion.

Moreover , Bush *et al.* (1995) extended this classification scheme of 1989 (Bush, 1989a;Bush,1989b and Bush,1989c) and attempted to correlate the functional characteristics with the molecular structure recognizing four major β - lactamase classes , one of which (Group2) is split into eight subgroups. Recently, Bush and Jacoby (2010) updated his classification scheme of 1995 by adding new functional subgroups to the scheme as a result of identification of new major β - lactamase families variants.

2.5.The Clinically Most Important β - Lactamases

2.5.1.Extended – Spectrum β - Lactamases (ESBLs)

ESBLs are mutant , plasmid mediated enzymes which derived from the older, broad spectrum β -lactamases and able to hydrolyze penicillins, first -, second cephalosporin and also third generation oxyimino cephalosporins and monobactams (Bali *et al.*, 2010 and Basavaraj *et al.*, 2011), but they are not active against cephamycins and carbapenems, generally they are inhibited by clavulanic acid , tazobactam and sulbactam (AL- Jasser , 2006).

ESBLs have evolved from point mutations, thus altering the configuration of the active site of the original and long known β - lactamases (Bradford,2001). Another important epidemiological characteristic is the possibility of these enzymes being transferred to other species via plasmid or transposons (Chaves *et al.*, 2011 and Ghafourian *et al.*,2011).

ESBLs are undergoing continuous mutation, causing the development of new enzymes showing expanded substrate profiles. At present, there are more than 300 different ESBL variants , and these have been clustered into nine different structural and evolutionary families based on amino acid sequence. TEM and Sulphhydryl variable SHV were the major types (Bali *et al.*, 2010). However, the new types of CTX-M were discovered later especially in *E.coli* and *K. pneumoniae* and appeared to be the most prevalence ESBLs during the last 5 years (Livermore,2007;Pitout and Laupland ,2008 and Jones *et al.*,2009).

a) TEM- β - lactamase

The TEM type ESBLs are derivatives of TEM-1 and TEM-2, TEM-1 is the most prevalence in this group (Poirel *et al.*, 2010). It was reported in 1965 from an *E. coli* isolate from a patient in Athens , Greece named Temoneira, hence named as TEM (Medeiros,1984).

TEM-1 hydrolyzes ampicillin at a greater rate than carbenicillin , oxacillin or cephalothin and has negligible activity against extended spectrum cephalosporins and they are inhibited by clavulanic acid (Upadhyay *et al.*,2010). TEM-2 has the same hydrolytic profiles as TEM-1 , but differs from TEM-1 by having single amino acid substitutions. TEM-3 is the first TEM type β - lactamase which show phenotypic characteristic of ESBLs. Although TEM- type ESBLs are presented mostly in *E. coli* and *K. pneumoniae*, they are also been found in other species of Gram- negative bacteria such as *Enterobacter aerogenes* , *Morganella morgannii*, *Proteus mirabilis*, *P. rettgeri* and *Salmonella* spp. with increasing frequency (Livermore, 1995 and Thirapanmethee, 2012).

b) SHV β - lactamase

The SHV-1 has approximately 68% amino acid sequence homology to TEM-1. They also have structure similarly . The SHV-1 β - lactamase is often found in *K. pneumoniae* and is involved in ampicillin resistance (Thirapanmethee, 2012). Unlike the TEM-type β -lactamases, there are relatively few derivatives of SHV-1. The majority of SHV variants possessing an ESBL phenotype are characterized by the substitution of a serine for glycine at position 238. Also some have a substitution of lysine for glutamate at position 240 (AL- Jasser, 2006).

c) CTX-M β - lactamase

The CTX-Ms are a more recent family of plasmid – mediated ESBLs which were initially identified by preferential hydrolysis of cefotaxime. They are inhibited better by the β - lactamase inhibitor tazohactam than by sulbactam and clavulanate (Bush *et al*.,1993 and Bradford,2001). Rather than arising by mutation , they represent examples of plasmid acquisition of beta- lactamase genes that are normally found on the chromosome of non- pathogenic bacteria genus *Kluyvera* especially *K. ascorbata* and *K. georgiana* (Bonnet , 2004).

The name of this group is derived from the fact that it is resistant to cefotaxime more than other oxyimino β - lactams such as ceftazidime, ceftriaxone and cefepime. This enzyme shows only 40% homology to TEM and SHV, and can be classified by amino acid sequences into 5 subgroups; CTX-M$_1$, CTX-M$_2$, CTX-M$_8$, CTX-M$_9$ and CTX-M$_{25}$ (Thirapanmethee, 2012).

d) OXA β - lactamase

The OXA – type enzymes are another growing family of ESBLs. These β - lactamases differ from the TEM and SHV enzymes in that they belong to molecular class D and functional group 2d (Bush *et al.*, 1995). They are resistant to ampicillin, cephalothin , oxacillin, and cloxacillin, but are not inhibited by clavulanic acid (Thirapanmethee, 2012).

Although , most ESBLs have been found in *E. coli* , *K. pneumoniae*, and other *Enterobacteriaceae*, the OXA- type ESBLs were originally discovered in *Pseudomonas aeruginosa* isolates from a single hospital in Ankara, Turkey (Hall *et al.*, 1993).

e) PER- β - lactamase

The PER- type ESBLs have about 25-27% amino acid homology with the known TEM and SHV type ESBLs. PER-1 β - lactamase hydrolyzes penicillins and cephalosporins efficiently , but is susceptible to clavulanic acid inhibition. The PER-1 β - lactamase was first detected in strains of *P. aeruginosa*, later, it was found among isolates of *S. enterica* serovar *Typhimurium* and *Acinetobacter* spp. The PER- type ESBL can be found worldwide, but is most frequent in Europe (Thirapanmethee , 2012).

f) VEB and GES β - lactamase

The prevalence of VEB and GES is rare compared with TEM , SHV, and CTX-M β - lactamase . They are found in limited geographic region, VEB-1 and VEB-2 isolated from south East Asia or GES-1 and GES-2 isolated from South Afirica, France and Greece (Thirapanmethee, 2012).

VEB-1 was first found in a single isolate of *E. coli* in Vietnam (Poirl *et al.*,1999). An identical β - lactamase has also been found in

K.pneumoniae,E.cloacae and *P. aeruginosa* in Thailand (Girlich *et al*.,2001) . VEB-1 has greatest homology to PER-1 and RER-2 (38%). It confers high level of resistance to ceftazidime , cefotaxime and aztreonam, which is reversed by clavulanic acid (Al- Jasser , 2006). Enzymes of GES are one of the families of class A ESBLs, the first producer isolate was identified in 1988 in France but originated from Guiana (Poirel *et al* 2000). They are encoded by gene cassettes present in integrons , observed in *P. aeruginosa* and several enterobacterial species (Gniadkowski, 2008). Despite their relatively low affinity , these enzymes often confer clear resistance to various β - lactam in clinical isolates (Poirel *et al* ., 2000).

2.5.2.AmpC β - lactamases

The first bacterial enzyme reported to hydrolyze penicillin was the AmpC β - lactamase of *E.coli* , although it was not given this name in the 1940s (Jacoby, 2009). AmpC β - lactamases are class C or group I cephalosporinases that confer resistance to a wide variety of β - lactam antibiotics including alpha methoxy β - lactams such as cefoxitin, narrow and broad spectrum cephalosporins , aztreonam, and are poorly inhibited by β - lactamase inhibitors such as clavulanic acid, sulbactam and tazobactam but may be inhibited by cloxacillin or boronic acid (Bush *et al*., 1995). Cefoxitin resistance is used as an indicator for AmpC-mediated resistance but it can also be an indication of loss outer membrane permeability (Philippon *et al*., 2002).

AmpC β - lactamases are of two types , chromosomal inducible and plasmid mediated non- inducible (Mohamudha *et al*., 2012). Chromosome-mediated AmpC β - lactamases have been described in a wide variety of Gram- negative bacilli (Mohamudha *et al*., 2010).

Overproduction of their chromosomal AmpC β - lactamase by mutation is probably responsible for the resistance in these organisms (Yan *et al.*, 2002). In most genera of the family *Enterobacteriaceae* , AmpC is inducible (Bush *et al.*, 1995). Plasmid mediated AmpC β -lactamase (PMABLs) have evolved by the movement of chromosomal genes on to plasmids and are found in *E. coli, K. pneumoniae, Salmonella* spp. , *P. mirabilis, Citrobacter freundii, E. aerogenes* which confer resistance similar to their chromosomal counterparts (Mohamudha *et al.*, 2012).

Unlike chromosome- mediated AmpC, most plasmid – mediated AmpC genes, such MIR-1, are expressed constitutively even in the presence of a complete system for induction (Phillipon *et al.*, 2002).

2.5.3. Carbapenemases

Carbapenems have the broadest spectrum of all β-lactam antibiotics and are increasingly used to treat infections caused by otherwise multidrug –resistant Gram-negative bacteria. Consequently, emerging resistance to carbapenems is the major public health concern ,especially when it involves acquired ,horizontally transmissible carbapenemases (Queenan and Bush, 2007). Those that hudrolyse imipenem and /or meropenem are classified in either Ambler classes A, B or D (genetic differences) or in Bush –Jacoby- Medieros groups 2f,3a or 3b (substrate preference and molecular structure).

Carbapenemases are enzymes that inactivate carbapenems and sometimes other classes of β - lactams. β - lactamase inhibitors such as clavulanic acid , tazobactam, and sulbactam, are inactive against carbapenemases. They are found in *Enterobacteriaceae, Pseudomonas* spp., and *Acinetobacter* spp. The majority of genes controlling carbapenemase production are transferable by plasmid (Marsik and Nambiar,2011).

2.5.3.1.Molecular Class A Carbapenemases

The molecular class A carbapenemases , including the plasmid-mediated serine β - lactamases KPC (for *K. pnenmoniae* carbapenemase) and GES (Guiana extended spectrum) and the chromosomally encoded SME (for *Serratia marcescens* enzyme) and IMI/ NMC (imipenem-hydrolyzing β - lactamase/ not metalloenzyme carbapenemase) enzymes, are effective carbapenemases (Forssten, 2009).

The hydrolytic mechanism of these enzymes requires an active- site serine at amino acid position 70 according to Ambler numbering system and presence of a disulfide bond between Cys69 and Cys238 (changes the overall shape of the active site). The structural changes decrease the steric hindrance caused by the C-6 hydroxyethyl side chain of carbapenems, resulting in increased hydrolysis of imipenem (Ambler *et al.*,1991and Queenan and Bush, 2007).

The class A carbapenemases have the ability to hydrolyze a broad spectrum of antibiotic , including penicillins, early and extended – spectrum cephalosporins, aztreonam as well as carbapenems, and all are inhibited by clavulanic acid and tazobactam (Queenan and Bush, 2007).

KPC stands for *K. pneumoniae* carbapenemase, this plasmid-mediated enzyme was first detected in a *K. pneumoniae* clinical isolate from North Carolina in 1996 (Yigit *et al.*, 2001). It is classified as Bush subgroup 2f class A serine – based enzymes, inhibited by clavulanic acid and tazobactam (Robledo *et al.*, 2011).

This carbapenemase – encoding gene is found on transferable plasmid associated with Tn3 – type transposon Tn 4401 (Arnold *et al.*,

2011). Ten KPC variants (KPC-2 to -11) have been described so far, differing between them in 1-or 2 – point mutation.KPC-2 and KPC-3 are the most common in clinical specimens and account for most epidemic outbreaks (Nordmann *et al.*,2009 and Chen *et al* .,2012).

Though mainly found in *K. pneumoniae* , there have been reports of these enzymes in other genera of the *Enterobacteriaceae* family, such as *K.oxytoca,Escherichia coli*, *Enterobacter* spp., *Serratia marcescens*, *Salmonella enterica,Protus mirabilis* and *Citrobacter freundi*. Worse still, KPC resistance has been reported in inherently- resistant organisms such as *A.baumannii* and *Pseudomonas* spp. (Villegas *et al.*,2007;Sacha *et al.*,2009 and Arnold *et al.*,2011).In terms of resistance, isolate producing KPC are resistant to all β -lactam agents including penicillins , early and late generation cephalosporins, cephamycins, aztreonam, carbapenems and β -lactam / β - lactamase inhibitor combinations (Patel *et al.*,2009).

Enzymes of the GES type also called IBC (integron borne cephalosporinase), is an infrequently encountered family that was first described in *K.pneumoniae* in Guiana (Poirel *et al.*, 2000) and in *E. cloacae* isolate in Greece (Giakkoupi *et al.*, 2000).

The chromosomal β - lactamases SME and IMI/NMC enzymes have been detected in rare clinical isolates among *Serratia marcecens* and *E. cloacae* isolates (Queenan and Bush, 2007).

2.5.3.2 Molecular Class B Metallo β - Lactamases

The molecular class B enzymes " metallo- β -lactamases" (MBLs) were first to distinguish from serine β - lactamases in 1980 (Ambler, 1980). Metallo -β - lactamases depend on heavy metals like Zn^{++} for β – lactam hydrolysis, due to this zinc dependency, chelators such as ethylenediamine tetra- acetic acid (EDTA) inhibit MBL activity (Marsik and Nambiar , 2011). They are resistant to well- known β - lactamase inhibitors like clavulanic acid , sulbactam, and tazobactam and confer resistance to all β - lactam antibiotics except monobactam (Mirsa, 2012).

The metallo β - lactamases are subdivided on the basis of sequence alignments into three subclass B1, B2 and B3. Phylogenetic studies suggest that B1 and B2 descend from a common ancestor and subclass B3 share only in structural similarities with these subclasses .B1 and B3 are able to bind one or two zinc ions , B2 are mono- Zn enzymes that have evolved specificity toward carbapenems. Of greater importance are the acquired or transferable families of MBLs which include IMP (active on imipenem), VIM (Verona integron – encoded metallo β - lactamase) , GIM (German imipenemase), SIM (Seoul imipenemase), SPM (Sao Paulo MBL) and NDM (New Delhei Metallo β - lactamase) which are located within gene cassettes as a part of integron structures (Walsh *et al.*,2005;Queenan and Bush, 2007 ;Dugal and Fernandes, 2011 and Mirsa,2012). These integron structures may then associate with transposons and plasmids which then can be easily transferred between bacteria (Tailor , 2011).

The IMP type of MBLs is a transferable type of beta-lactamase which confers the property of hydrolyzing imipenems as well as some extended spectrum cephalosporins. They are insensitive to most inhibitory agents but are however susceptible to aztreonam (Dugal and Fernandes, 2011). The first indication of the IMP type was found in the Japanese region in a *P. aeurginosa* strain GN17203 in 1988 (Watanabe *et al.*,1991). They were subsequently reported in four *S. marcescens* isolates in Japan.

IMP-1 was first detected in *K.pneumoniae* DB96 in 1999 when a clinical isolate was analyzed from a blood sample of a leukemic patient in

Singapore(Koh *et al.*,2001). The IMP genes are located on transferable conjugated plasmid of about 120KD which could be readily mobilized to other Pseudomonas strains. This same gene was then transferred to four *S. marcescens* collected 32 in seven general hospitals in Japan in 1993 (Dugal and Fernandes, 2011).

VIM class 1 integron associated MBL was first observed in a *P. aeruginosa* isolate in Verona , Italy in 1997 (Cornaglia *et al.*,2000). It is closely related to BCII from *Bacillus cereus* sharing 39% amino acid identity (Yong *et al.*,2009)

These types are mainly borne on integrons on plasmid, they are the second most dominant group of transferable MBLs (Naas, 2010). VIM enzymes are resistant to a number of beta lactams like piperacillin, cefazidime, imipenem and aztreonam. These enzymes are dependent on metal ions, which is indicated by loss of activity on addition of EDTA and restoration up on addition of Zn^{++} (Duga and Fernandes, 2011).

GIM-1 was isolated in Germany in 2002 from *P. aerugenosa* strain (Castanheira *et al.*, 2004). The enzyme SIM was first detected in *A baumannii* strains in Korea. SIM-1 exhibited (64 to 69 %) amino acid identity to the IMP enzymes (Lee *et al.*, 2005). SPM was first detected from the clinical *P. aerugenosa* strain in 1997 in Brazil and designated *bla* SMP-1 (Toleman *et al.*, 2002). Since their initial discoveries, SPM , GIM and SIM metallo β -lactamases have not spread beyond their countries of origin (Queenan and Bush, 2007) .

2.5.3.2.1 New Delhi Metallo β-Lactamase (NDM)

In 2009, a novel MBL subtype was recognized in a *K. pneumoniae* isolate obtained from a Swedish patient of Indian origin , who had received medical care in New Delhi, India, hence named NDM-1 (New Delhi MBL-1) (Marsik and Nambiar , 2011).

NDM-1 expressed high level of resistance to all β - lactam antibiotics but remained susceptible only to colistin and tigecycline. It shares very little identity with other metallo β - lactamases, maximum identity has been observed to VIM-1 /VIM-2 (32.4%). Compared to VIM-2, NDM-1 displayed tight binding to most cephalosporins and , in particular to cefuraxime cefotaxime, and cephalothin and also to the penicillins (Shakil *et al.*, 2011).However, it does not bind to carbapenem as tightly as IMP-1 or VIM-2 (Yong *et al.*, 2009).

2.5.3.3.Molecular Class D OXA Carbapenemases

Class D enzymes are OXA (for oxacillin hydrolyzing) enzymes, which are penicillinases capable of hydrolyzing oxacillin and cloxacillin. These serine beta- lactamases are plasmid encoded and are found primarily in *P. aeruginosa*, *A. baumannii* ,and rarely in isolates of *Enterobacteriaceae* from the United States. The major concern with OXA carbapenemases is their ability to rapidly mutate and expand their spectrum of activity (Marsik and Nambiar,2011). .

The first OXA enzyme with carbapenemase activity was found in an isolate of *A. baumannii* in 1985 from a patient in Scotland. This enzyme was called ARI-1(for *Acinetobacter* resistant to imipenem) which was later renamed OXA-23 (Paton *et al* .,1993).The hydrolysis of carbapenems by the class D oxacillinase in *A. baumannii* was classified into four subgroups of eight clusters and they have been designated OXA-23-like;OXA-40-like;OXA-51-like and OXA-58-like (Brown and Amyes,2006). Whereas Walther-Rasmussen and Hoiby (2006) subclassified carbapenem-hydrolysing OXA enzymes into eight distinct branches or subgroups including OXA-23, OXA-24, OXA-48, OXA-50, OXA-51, OXA-55, OXA-58 and OXA-60. Another subgroup, OXA-62 was identified as species-specific oxacillinase in *Pandoraea pnomemusa* from cystic fibrosis patients (Schneider *et al.*,2006). OXA-51-like enzymes are intrinsic and naturally found in all *A. baumannii* strains tested (Heritier *et al.*,2005).

Among the carbapenem hydrolyzing OXA enzymes, there is 40-70% amino acid homology within groups and within the group the homology is 92.5% or higher. The catalytic mechanism of these enzymes is similar to that of serine carbapenemases. These enzymes have measurable activity against penicillins, early cephalosporins and imipenem (Walther-Rasmussen and Hoiby ,2006 and Queenan and Bush, 2007).

2.5.3.4. Detection of Carbapenemases

The first indication to identify carbapenemases production in a clinical isolate is an observed reduced susceptibility of carbapenem with MIC values up to 2 µg /ml. In many previous studies , scientists noted that transfer methods from strains harboring the genes responsible for metallo – β - lactamases into *E.coli* , resulted in much lower MIC of the recipient strain than that of the parent (Queenan and Bush, 2007). This demonstrates the presence of other mechanisms of resistance like decreased permeability (Cao *et al.*,2000).Elevated carbapenems MICs is an indicator that predicts carbapenmases production in *Enterobacteriaceae* , but it is rare to encounter high MIC values above the CLSI breakpoints of resistance (Queenan and Bush, 2007). The KPC serine carbapenemases also have been reported to be difficult to detect (Bratu *et al* ., 2005 b and Moland *et al.*, 2003). They are often associated with imipenem MICs as low 2 µg /ml (Moland *et al.*, 2003) , and a low inoculum has resulted in susceptible MICs by broth microdilution (Bratu *et al.*, 2005a).

Furthermore , frequent problems are faced by laboratories with inconsistent E- test results that were due to colonies present in the zones of inhibition , this might be described by mutants being located above, close or lower than the susceptibility zone of inhibition (Queenan and Bush, 2007).

Using β - lactamases inhibitors with combination of antibiotics , based on a disk method or even on E test strips , are counted as a phenotypic test used for screening for carbapenemases production . The test is interpreted as positive when an increase of the zone of inhibition(for the disk), or a decrease in the MIC is seen in comparison with the antibiotic alone. Clavulanic acid is usually used for the detection of a class A β - lactamases whereas testing with EDTA is often used as a screen for metallo β - lactamases producers (Arakawa *et al.*,2000). When the presence of a carbapenmase is suspected , PCR is the gold standard method and the fastest way to determine which family of β - lactamases is present (Queenan and Bush, 2007).

2.5.3.5.Epidemiology

International travel is an important risk factor for colonization or infection with antibiotic- resistant organisms, the risk being highest among travelers to India, the Middle East and Africa (Munoz-Price and Quinn,2009;Poriel *et al.*,2011and El-Herte *et al.*,2012).

Genes coding for KPC carbapenemases are all coded on plasmids that also associated with resistance determinants for other antibiotics (Arnold *et al.*,2011). KPCs have been reported in the United States , Europe, Asia and Australasia (Chen *et al.*, 2012). Several instances of outbreaks and transmission of KPC-producing organisms were reported, predominantly from the northeastern United States. In 2002 – 2003 surveillance study in New York City , 9 of 602 *K.pneumoniae* isolates were found to contain the *bla*KPC gene. In the following year, 20 additional KPC- producing isolates were identified from 2 hospital outbreaks in the city (Bratu *et al.*, 2005a).

Although rare , GES enzymes have been identified worldwide including Greece, France , South Africa, Portugal, Brazil , Argentina, Korea and Japan . Most of the isolations were of single occurrences, however , one clone of *P. aeruginosa* was associated with a small hospital outbreak in South Africa during 2000 (Porile *et al.*,2002 and Queenan and Bush, 2007). Isolates producing GES enzymes with carbapenemase activity have been observed predominantly in Europe, South Africa and China (Queenan and Bush, 2007).

IMP- type carbapenemases have been reported from China, Taiwan , Korea and Australia , but have not spread extensively to other regions of the world , while VIM- type carbapenemase continue to spread over the world (Gupta *et al.*,2011and Kosmidis *et al.*, 2012) , GIM , SIM and SPM metallo β - lactamase have not spread beyond their countries of origin.

VIM-2 is the most commonly isolated MBL from many parts of the world , but since 2009, reports of NDM MBL detections are on rise (Marsik and Nambiar , 2011). One study reported 37 NDM-1 producing isolates from the UK , 44 isolates from Chennai , South India, 26 isolates from Haryana, North India and 73 isolates from various other sites in the Indian subcontinent (Kumarasamy *et al* ., 2010) . In 2010 itself, Canada, Japan and Sultanate of Oman also confirmed their first cases of NDM-1 harboring pathogens (Tijet *et al.*, 2011; Chihara *et al.*, 2011 and Poirel *et al.*, 2011). Subsequently , the first death due to infection by NDM-1 producing bacteria was reported in August,2010 (Sharma *et al.*, 2010).

OXA-48 is mainly found in *K. pneumoniae* and has been reported from Egypt, Turkey, China, India, and United Kingdom (Young *et al.*,

2009 and Poirel *et al.*,2013).OXA-24 and OXA-40 were reported from hospital outbreaks by *A. baumannii* in Spain and Portugal.OXA-23 has been reported from hospital outbreaks by *Acinetobacter* spp. In Brazil, UK and Korea.OXA-23 and OXA-58 have been recovered from infections in civil and military personnel serving in Iraq and Afghanistan (Hujer *et al.*,2006).OXA-58 is routinely found in *Acinetobacter* spp isolates in France, Greece, Romania, Italy, Turkey, Kuwait and Argentina (Queenan and Bush, 2007).

2.6.Treatment of Carbapenemase- producing *K.pneumoniae*

Unfortunately , Gram- negative bacteria are becoming increasingly non- susceptible to carbapenems, due to the acquisition of new resistance mechanisms, largely attributed to the widespread use of these antimicrobials. Often, concurrent carriage of additional resistance determinants leads to decreased susceptibility to other classes of antimicrobials including quinolones and aminoglycosides. Therefore, therapeutic options for infections caused by carbapenem resistance bacteria are extremely limited and there are no established guidelines for their management (Boucher *et al.*,2009 and Kosmidis *et al.*, 2012).

However, combination therapy of different antibiotics may improve the chances of cure in highly resistant infections. Bulik and Nicolau, (2011) showed that the combination of doripenem plus ertapenem demonstrated enhanced efficacy over either agent alone against KPC producer.

The best currently available options against *K. pneumoniae* infections seem to be colistin, tigecycline, fosfomycin and aminoglycosides (Kosmidis *et al.*, 2012). Caution must be taken within these agents as colistin has neurotoxic and nephrotoxic effects and tigecyclines demonstrates low serum and urine concentrations, beside emergence of fosfomycin and aminoglycosides resistance which make these less appealing options for use (Nordmann *et al* .,2009;Neuner *et al.*,2011and Kosmidis *et al.*, 2012).

Several novel agents currently in development and have shown good *in vitro* activity against carbapenemase producing organisms. Avibactam (NXL-104), is a carbapenemase inhibitor which has shown excellent activity against β - lactamase producing *Enterobacteriaceae*, including KPC- types and OXA- 48 (Mushtaq *et al.*, 2010; Lagce- Wiens *et al.*, 2011 and Livermore *et al.*, 2011). Plazomicin (ACHN- 490), a new aminoglycoside, NAB739 and NAB7061 which are experimental polymyxin derivatives that are potentially less nephrotoxic than colistin, may be a viable options along with the novel monobactam, BAL30072 (Patel and Bonomo, 2011). Another agent , BAL30376 is a monobactam-clavulanic acid combination with activity against MBLs (Livermore *et al.*, 2010).

2.7.Mechanisms of Resistance Transfer

Transferable β - lactamase genes can be spread from a donor to a recipient on plasmids , transposons , insertion sequences and integrons, by conjugation , transduction or transformation. The result cell has a genome different from the donor or recipient (Courvalin , 2006).

a) Conjugation

This is a process of unilateral transfer of genetic material among bacteria of the same or different species. It involves direct cell to cell contact among bacteria. It involves the transfer of R- plasmid through a cytoplasmic bridge from a donor cell to a recipient cell. It is mediated by a plasmid or a transposon. It is a common phenomenon among different genera of Gram- negative bacteria. Transfer of plasmid also occurs among Gram- positive bacteria but there are essential differences in the mechanisms used to establish cell to cell contact. They do not appear to use pili as an initiator of conjugation, instead transfer appears to be by aggregation of bacterial cells mediated by cell surface structures (Grohmann *et al.*, 2003 and Elufisan *et al.*, 2012).

b)Transformation

Transformation is a process by which bacteria pick up pieces of a naked DNA from an environment and add them up to their own chromosome. The process may occur spontaneously in which the bacterial chromosome leak out from the donor cell to the recipient cell or it may occur by an artificial means as a result of extraction by chemical procedures. It may also be due to cellular break after lysis by bacteriophages. Transfer only takes place between competent cells. Transfer occurs only in bacteria and it is commonly found in Gram positive bacteria which are capable of taking up high molecular weight DNA from the aqueous environment (Elufisan *et al.*, 2012).

c) Transduction

Transduction is a process in which the phage particles are packaged with bacterial DNA instead of phage (van-Hoek *et al.*, 2011). Bacteriophages are extrachromosomal genetic elements (DNA, RNA) termed bacterial viruses due to their ability to infect bacterial cells and to transfer independently. To enter the host cell, the bacteriophage must attach to specific receptors on the surface of the bacteria and may therefore only infect bacteria carrying these receptors. Once inside the cell they may integrate into the host genome without killing the host or replicate in large numbers causing the cell to lyse (McDermott *et al.*, 2003).

2.8. Horizontal Gene Transfer of Resistance Genes

Several types of mobile genetic elements have been described to date, which play an important role in acquisition, maintenance, and spread of antimicrobial resistance genes. In this regard, plasmids, integrons and transposons are the most important elements (van-Hoek *et al.*, 2011 and Stalder *et al.*, 2012)

a) Plasmids

Plasmids also known as R- factor , are extra chromosomal elements which contain their own origin of replication and confer resistance on bacteria to antibiotics (van -Hoek *et al.*, 2011 and Elufisan *et al.*, 2012). Plasmids , are like chromosomes because they are capable of independent replication but they differ from bacterial chromosomes because they are smaller with an approximate size of between 0.1 to 10% of the chromosome in bacteria (Elufisan *et al.*, 2012). Most are circular, double-stranded DNA molecules, which encode up to 10% of the host cell chromosome (Giedraitiene *et al.*, 2011). Plasmid may be conjugative (self- transmissible) or nonconjugative (unable to cause their own transfer) , some plasmids have a broad host range and can transfer between different species whereas others have a much narrower host range and are confined to one genus or species (Rice , 2000 and van-Hoek *et al.*, 2011).

Plasmids do not play any role in the normal functioning of bacterial cell but may confer on the bacterial character that can give the bacteria some survival advantage in certain condition such as the ability to survive in an unfavorable condition like in the presence of an antibiotic. Plasmid may also code for other properties in bacteria such as the ability to produce toxin, utilize or ferment unusual sugar or food source as camphor, production of pili for the attachment of a cell to substrate (Elufisan *et al.*, 2012).

b) Integrons

Integrons are gene capture system found in plasmids, chromosomes and transposons. They recognize and capture multiple gene cassettes. A gene cassette may encode genes for antibiotic resistance, although, most genes in integrons are uncharacterized. Therefore, integrons have been identified as a primary source of resistant genes within microbial population and were suspected to serve as reservoir of antimicrobial resistance gene within microbial population (Elufisan *et al.*, 2012 and Stalder *et al.*, 2012).

The cassette has a specific recombination sites that confer mobility because it is recognized by recombinase encoded by the integron that catalyses its integration into specific site within the integrons (Giedraitiene *et al.*,2011).

Two major groups of integrons have been described chromosomal integrons (CIs) , and mobile integrons (MIs). CIs are located on the chromosome of hundreds of bacterial species (Cambray *et al.*, 2010). CIs have also been termed super- integrons (SIs) as they can carry up to 200 cassettes that mainly encode proteins with unknown functions. MIs are not self- transposable elements but are located on mobile genetic elements such as transposons and plasmids, which promote their dissemination among bacteria (Stalder *et al.*, 2012).

c)Transposons

The discovery of transposons (or transposable element) often referred to as jumping gene has provided explanatation for frequent occurrence of drug resistance in bacteria (Elufisan *et al.*, 2012). Transposons are units of DNA that move from one molecule of DNA to another inserting themselves nearly at random by recombination, which is mediated by a transposase encoded within the transposon (Thomas , 2007). They differ from plasmids in that although some from a circularized intermediate, they do not replicate independently and are usually incorporated into the bacterial chromosome (Mullany *et al.*, 2002).

3.Materials and Methods

3.1. Materials

3.1.1. Equipments and Instruments

Type of equipment	Manufacture (Origin)
Autoclave	Hiclave-Hirayama (Japan)
Centrifuge	Labner (Taiwan)
Cold centrifuge	Hermile (Germany)
Compound light microscope	Olympus (Japan)
Deep freezer	Froilabo (France)
Digital camera	Sony (Japan)
Distillator	GFL (Germany)
Electric oven	Memmert (Germany)
Electrophoresis unit	Auto (Japan)
Gel documentation system	Optima
Incubator	Binder (Germany)
Laminor flow cabinets	Labogene (Denmark)
Micropipette set 0.5-1000 µl	Eppendrof (Germany)
Millipore filter 0.22 mm	Difco (USA)
Nano drop	Thermo (USA)
PCR system	Labnet (USA)
pH – meter	LKB (Sweden)
Sensitive balance	Sartorius (Germany)
Shaker incubator	Memmert
Spectrophotometer	Shimadzu (Japan)
UV- Transilluminator	Optima
VITEK	BioMerieux (France)
Vortex	Thermolyne (USA)

Water bath	GFL

3.1.2. Biological and Chemical Materials

Biological and Chemical type	Manufacture (Origin)
Agar- Agar	Himedia (India)
Agarose	Bio basic(USA)
Barium chloride dihydrate $(BaCl_2).2H_2O$	Fluka (Switzerland)
Cellobiose	BDH (England)
Chloroform	Qualikems (India)
Disodium hydrogen phosphate (Na_2HPO_4)	BDH
DL- phenylalanine	BDH
D- Mannitol	BDH
Deionized sterile distilled water	Bioneer (Korea)
D- Sorbitol	BDH
D- Xylose	BDH
DNA Loading dye	Geneaid (Korea)
ESBL supplement	CHROMagar (France)
Esculin $(C_{15}H_{16}O_4.15H_2O)$	Himedia
Ethanol absolute	GCC (England)
Ethidium bromide solution	Bio basic
Ethylene diaminotetracitic acid (EDTA) $(C_{10}H_{14}N_2Na_2O_8.2H_2O)$	Thomas Baker (India)
Ferric chloride $(FeCl_3)$	Panreac (Spain)
Gelatin	Barcelana (Spain)
Glucose $(C_6H_{12}O_6)$	Panreac
Glycerol $(C_3H_8O_3)$	GCC

HgCl$_2$	Sigma (USA)
Gram stain	Himedia
Hydrochloric acid (HCl)	BDH
Hydrogen peroxide (H$_2$O$_2$) 6%	SDF (Iraq)
India ink	BDH
Iodine	GCC
Isoamyl alcohol	Qualikems
Isopropylalcohol	Mast Diagnostic (USA)
Kovac's reagent	Himedia
KPC supplement	CHROMagar
Lactose	BDH
L- Arabinose	Fluka
L-Rhaminose	BDH
Methyl red	Himedia
α - napnthol (C$_{10}$H$_8$O)	BDH
Nuclease free water	Promega (USA)
Peptone	Difco (USA)
Phenol	Panreac
Phenol red	Himedia
Potassium dihydrogen phosphate (KH$_2$PO$_4$)	BDH
Potassium hydroxide (KOH)	Panreac
Potassium iodide (KI)	GCC
Raffinose	Rediel dehiun (Germany)
Salicin	BDH
Sodium carbonate (Na$_2$HCO$_3$)	BDH
Sodium chloride (NaCl)	BDH
Sodium dihydrogen phosphate (NaH$_2$PO$_4$.	BDH

$2H_2O$)	
Sodium dodecyl sulfate (SDS)	Panreac
Sodium hydroxide (NaOH)	BDH
Starch soluble	Difco
Sulfuric acid (H_2SO_4)	Difco
Sucrose ($C_{12}H_{22}O_{11}$)	Difco
Tetramethyl *p* - phenylenediamine dihydrochloride	BDH
Tris- EDTA (TE) buffer	Bio basic
Tris- Borate – EDTA buffer (TBE buffer)	Bio basic
Tris- (hydroxymethyl) methylamine- NH_2-($CH_2OH)_3$ (Tris-OH)	BDH
Trehalose dihydrate ($C_{12}H_{22}O_{11}.2H_2O$)	Himedia
Typtone	Biolife (Italy)
Urea	BDH
Yeast extract	Himedia

3.1.3. Culture Media

Medium	Manufacturer (Origin)
Amyes transport medium	Himedia (India)
Blood agar base	Himedia
Brian- heart infusion agar	Himedia
Brian- heart infusion broth	Himedia
CHROMagar orientation	CHROMagar (France)
Eosin methylen blue agar	Biolife (Italy)
Kligler's iron agar	Oxoid (England)
Luria-Bertani broth	Himedia
MacConkey's agar	Himedia
Malonate broth	Difco (USA)
Moeller decarboxylase broth	Himedia
MR- VP broth	Himedia
Mueller- Hinton agar	Oxoid
Nutrient agar	Himedia
Nutrient broth	Himedia
Peptone water	Biolife
Simmons citrate agar	Himedia
Tryptic soy broth	Alfa (USA)
Tryptone soya broth	Oxoid
Urea agar base	Himedia

3.1.4. Antibiotics
3.1.4.1. Antibiotic disks

Antibiotic class	Antibiotic subclass	Agent used	Symbol	Content	Origin
Penicillin	Aminopenicillin	Ampicillin	AM	10 μg	Bioanalyase (Turkery)
	Ureidopenicillin	Piperacillin	PRL	100 μg	Bioanalyase
	Carboxypenicillin	Carbenicillin	PY	100 μg	Bioanalyase
β - lactams/ β - Lactamase inhibitor combination		Amoxicillin-Clavulanic acid (Amoxi-clav)	AMC	30 μg (20 μg 10 μg)	Bioanalyase
Cephems (Oral)	Cephalosporin	Cefaclor	CF	10 μg	Himedia (India)
		Cefprozil	CPR	30 μg	Himedia
Cephems (Parenteral)	Cephalosporin III	Cefotaxime	CTX	30 μg	Bioanalyase
		Ceftazidime	CAZ	30 μg	Himedia
		Ceftriaxone	CRO	30 μg	Bioanalyase
	Cephalosporin IV	Cefepime	FEP	30 μg	Bioanalyase
	Cephamycin	Cefoxitin	FOX	30 μg	Bioanalyase
Monobactams		Aztreonam	ATM	30 μg	Bioanalyase
Penems	Carbapenem	Imipenem	IMP	10 μg	Bioanalyase
		Meropenem	MEM	10 μg	Bioanalyase
		Ertapenem	ETP	10 μg	Himedia
Aminoglycosides		Gentamicin	CN	10 μg	Bioanalyase
		Amikacin	AK	30 μg	Bioanalyase
		Kanamycin	K	30 μg	Bioanalyase
Quinolones	Quinolones	Nalidixic acid	NA	30 μg	Bioanalyase
	Fluoroquinolones	Ciprofloxacin	CIP	5 μg	Bioanalyase
		Levofloxacin	LE[5]	5 μg	Himedia
Folate pathway Inhibitors		Trimethoprim-Sulfamethoxazole	SXT	25 μg (1.25 μg)	Bioanalyase
Nitrofurons		Nitrofurantion	F	300 μg	Bioanalyase
Phenicols		Chloramphenicol	C	30 μg	Bioanalyase
Tetracyclines		Tetracycline	TE	30 μg	Bioanalyase
		Doxycycline	Do	30 μg	Bioanalyase

3.1.4.2. Antibiotic Powders

Antibiotic	Manufacturer (Origin)
Penicillin G	Fluka (England)
Ampicillin	Troge (Germany)
Amoxicillin	Asia (Syria)
Ceftazidime	Trolian (Spain)
Rifampicin	Ajanta (India)
Meropenem	Astrazeneca (Japan)

3.1.5. Kits

Kit type	Manufacturer (Origin)
VITEK 2 system	BioMerieux (France)
Genomic DNA Extraction	Geneaid (Korea)
HiComb MIC strip	Himedia (India)
Minimum inhibitory concentration evaluator	Oxoid (England)

3.1.6. Standard Bacterial Strain

Strain name	Laboratory identifier	Key Characteristics	Source
Escherichia coli	American Type Culture collection (ATCC 25922)	Wild type	University of Kufa, College of Medicine, Department of Microbiology.
Escherichia coli	MM294	end AI; hsd M+; hsd R-; sup E; pro $^-$; th $^-$; rif^r	University of Kufa, College of Medicine, Department of Microbiology

end AI: Endonuclease AI deficiency; hsd M+: Modification system not present; hsdR- : Rrstriction enzyme deficiency; sup E: Amber; pro $^-$: need to proline; th $^-$: need to thiamine; rif^r : resistance to rifampin

3.1.7. Polymerase Chain Reaction Materials

3.1.7.1. Master Mix

Go Taq® Green Master Mix,2X	Source
Go Taq ®DNA polymerase is supplied in 2x Green Go Taq® Reaction buffer (pH 8.5), 400 µM d ATP, 400 µM dGTP, 400 µM dCTP, 400 µM d TTP and 3mM MgCl$_2$.	Promega (USA)

3.1.7.2. Molecular Weight DNA Marker

100 bp DNA Ladder	100-2000 base pairs. The DNA ladder consists of 13 double strand DNA fragments ranging in sizes from 100 to 1,000 bp increments, and additional fragments of 1,200; 1, 600; 2,000 bp. The 500, 1,000 and 2,000bp bands are two to three times brighter for easy identification.	Bioneer (Korea)

3.1.7.3.β - Lactam Resistance Primers (Bioneer, Korea)

Type of β - lactamase	Primer name	Oligosequence (5′ 3′)	Pmoles	Quantity of D.D.W (µl) **	Product size (bp)	Reference
	KPC	F: TGTCAC TGT ATC GCC GTC TAG	15510	155.1	821	Munoz-Price *et al.*,2010
		R: TTACTGCCCGTTGACGCCCAATCC	14160	141.6		
	VIM	F: AGTGGTGAGTATCCGACAG	10270	102.7	261	Taskris *et al.* ,2000 Gorbner *et l.*, 2009
		R: ATGAAAGTGCGTGGAGAC	10660	106.6		
	IMP	F: GATGGTTTGGTGGTTCTTGT	11000	110	488	Sung *et al.*,2008 Jeon *et al.*, 2005
		R: ATAATTTGGCGGACTTTGGC	10510	105.1		
	OXA-23	F: TCTGGTTGTACGGTTTCAGC	11450	114.5	606	Srinivansan *et al.*,2009 Hujer *et al.*,2006
		R: AGTCTTTCCAAAAATTTTG	10900	109		
	SME	F: AACGGCTTCATTTTTGTTTAG	15280	152.8	830	Queenan *et al* .,2000
		R: GCTTCCGCAATAGTTTTATCA	11580	115.8		
	NDM-1	F: GGTTTGGCGATCTGGTTTTC	10700	107	621	Nordmann *et al.*,2011 Li *et al.*,2012
		R: CGGAATGGCTCATCACGATC	14000	140		
*ESBL	TEM	F: AAACGCTGGTGAAAGTA	11550	115.5	822	Paterson *et al* .,2003 Hujer *et al.*,2006
		R: AGCGATCTGTCTAT	14890	148.9		
	SHV	F: ATGCGTTATATTCGCCTGTG	10700	107	753	Paterson *et al* .,2003 Hujer *et al.*,2006
		R: TGCTTTGTTATTCGGGCCAA	10890	108.9		
	CTX-M	F: CGCTTTGCGATGTGCAG	12920	129.2	550	Paterson *et al* .,2003
		R: ACCGCGATCGTTGGT	12310	123.1		
	VEB	F: ACCAGATAGGAGTACAGACATATG	11120	111.2	727	Park *et al.*,2009
		R: TTCATCACCGCGATAAAGCAC	14840	148.4		
	PER	F: AGTCAGCGGCTTAGATA	11540	115.4	978	Wei-feng *et al.*,2009
		R: CGTATGAAAAGGACAATC	10610	106.1		
	GES	F: GCGTTTTGCAATGTGCTCAAC	15570	155.7	846	Hosoglu *et al.*,2007
		R: CGCCGCCATAGAGGACTTTAG	14860	148.6		
	OXA-1	F:ATATCTCTACTGTTGCATCTCC	15130	151.3	619	Karami *et al* .,2008
		R: AAACCCTTCAAACCATCC	11790	117.9		
AmpC β - lactamase	AmpC	F: ATCAAAACTGGCAGCCG	11940	119.4	550	Paterson *et al* .,2003 Kaczmarek *et al.*,2006
		R: GAGCCGTTTTATGCACCCA	10810	108.1		

* **ESBL**: Extended spectrum β - lactamase ; **represent quantity of deoinized distilled water to get 100 picomole/ µl as stock solution.

3.2. Methods

3.2.1. Preparation of Buffers and Solutions

The following solutions and buffers were used in the present study. Those, which require sterilization were autoclaved at 121C° for 15-20 min at 15 psi. Millipore filters (0.22 μm) were used for sterilization of heat –sensitive solutions like antibiotic ,urea and sugars. The pH of the solution was adjusted using 1M NaOH or 1M HCl.

3.2.1.1. McFarland's 0.5 Turbidity Standard

The 0.5 McFarland's solution was prepared by adding 0.5 ml of 1.175% (w/v) $BaCl_2.2H_2O$ solution to 99.5 ml of 1% (v/v) H_2SO_4. The McFarland standard tubes were sealed with parafilm to prevent evaporation and stored for up to 6 months in the dark at room temperature. The accuracy of the density of a prepared 0.5 McFarland's standard was checked by using a spectrophotometer. The absorbance of the wavelength of 625nm should be between 0.08 and 0.1 (CLSI, 2010).

3.2.1.2. Normal Saline Solution

It was prepared by dissolving 0.85 gm of NaCl in 90 ml of distilled water and further completed to 100 ml with D.W. autoclaved at 121 C° for 15min (Collee *et al.*, 1996).

3.2.1.3. Phosphate Buffer Solution(PBS)

This buffer consists of two solutions and was prepared as follows:

 Solution A: 3.12 gm of NaH_2PO_4. $2H_2O$ was dissolved in 90 ml of D.W. and then completed to 100ml with D.W.

 Solution B: 2.839 gm of Na_2HPO_4 was dissolved in 90 ml of D.W. and the volume was completed to 100 ml.

Then, 87.7 ml of solution A was added to 12.3 ml of solution B and the pH was adjusted to 6. The buffer was used for the detection of β - lactamase production (Collee *et al.*, 1996).

3.2.1.4. β - lactam antibiotic solutions

The solutions were prepared as stock solution with concentration of 10 mg/ml by dissolving 1gm of the antibiotic in a small volume of sterile PBS (pH 6.0) for ampicillin and amoxicillin. Each of these solutions was further diluted with sterile D.W. to volume 100 ml, and stored at 4 C° until being used (CLSI, 2010).

3.2.1.5. Solutions used for β -lactamase detection:

These solutions were prepared according to Collee *et al.* (1996) as follows:

Penicillin G solution: It was prepared by dissolving 0.569 gm of penicillin G in PBS (3.2.1.3.). The solution was sterilized, dispensed in small vials, and stored at-20 C°.

Starch solution: It was prepared by dissolving 1gm of soluble starch in 100 ml of D.W. and boiled in water bath for 10 minutes , and stored in a dark bottle at 4 C°.

Iodine solution: iodine (2.03 gm) and KI (5.32 gm) were dissolved in 90 ml of D.W. then the volume was completed to 100 ml with D.W., and stored in a dark bottle at 4 C°.

3.2.1.6. EDTA Solution for Disks Preparation

This solution was prepared by complete dissolving of 190 mg of EDTA in 1ml of D.W., pH was adjusted to 8, then it was sterilized by autoclaving. 10 ml of EDTA solution was added to a 6- mm Whatmann filter No.1 (n=100) and allowed to dry. Each disk contain approximately 1,900 μg of EDTA and used for detection of metallo-β -lactamases producing isolates (Lee *et al.*,2003).

3.2.1.7. CHROMagar and Supplement Solution

The supplements were prepared according to the manufacturer recommendations by dissolving 57mg/ml of ESBL supplement and

40mg/ml of KPC supplement, separately, in sterile D.W., vortexed, homogenized and added in proportions of 10 ml/l of final melted orientation CHROMagar after cooled at 45C°, then poured in plates and used freshly.

3.2.1.8. Solutions Used in DNA Extraction

The following solutions were prepared as described by Pospiech and Neuman (1995) with some modifications according to the current study:

3.2.1.8.1. Tris – EDTA (TE) Buffer

This buffer was prepared by adding 0.05M Tris-OH and 0.001M EDTA to 800 ml D.W. The pH was adjusted to 8 and completed to one liter by D.W., then autoclaved at 121 C° for 15 minutes, and stored at 4 C° until used.

3.2.1.8.2. Salt-EDTA-Tris (SET) Buffer

It was prepared by dissolving 20 mM Tris-OH, 25 mM of EDTA and 75 mM NaCl in 750 ml D.W., the pH was adjusted to 8 and the volume completed to 1000 ml by D.W., then autoclaved at 121 C° for 15 minutes.

3.2.1.8.3. Sodium Dodecyl Sulfate (SDS) Solution (25%)

Sodium dodecyl sulfate (25mg) was dissolved in 100 ml of D.W., then sterilized by autoclave at 121 C° for 15 min.

3.2.1.8.4. Sodium chloride (NaCl) solution (5 M)

Sodium chloride (14.625 gm) was dissolved in 50 ml D.W., sterilized by autoclave, and stored at 4 C°.

3.2.1.8.5. Phenol: Chloroform: Isoamyl Alcohol (25:24:1)

The solvent was composed from 25 ml phenol, 24 ml chloroform, and 1ml isoamyl alcohol.

3.2.2. Preparation of Reagents

The following reagents were prepared as described by MacFaddin (2000).

3.2.2.1. Oxidase Reagent

It was prepared in a dark bottle by dissolving 0.1gm of tetramethyl *p*- phenylenediamine dihydrochloride in 10 ml of D.W.

3.2.2.2. Catalase Reagent

Hydrogen peroxide (3%) was prepared from the stock solution and stored in a dark bottle. It was used for detecting the ability of bacteria to produce catalase enzyme.

3.2.2.3. Methyl Red Reagent

It was prepared by dissolving 0.1gm of methyl red in 300 ml of 95% ethylalcohol and then completed to 500 ml with D.W. This reagent was used as an indicator in methyl red test.

3.2.2.4. Voges- Proskauer Reagents

The reagents were prepared as follows:

Reagent A: α - naphthol (5%): is prepared by dissolving 5gm of α - naphthol in 100 ml of absolute ethanol.

Reagent B: KOH (40%): is prepared by dissolving 40 gm of KOH in 100 ml of D.W.

3.2.2.5. Gelatin Liquefaction Reagent

This reagent was prepared by dissolving 5 gm of $HgCl_2$ in 20 ml of concentrated HCl, then the mixture was completed to 100 ml with D.W. (Collee *et al.*,1996).

3.2.2.6. Aqueous Ferric Chloride Reagent

It was prepared by dissolving 12 gm of $FeCl_3$ in 2.5 ml of concentrated HCl and completed to 100 ml with D.W.

3.2.3. Preparation of Culture and Diagnostic Media

3.2.3.1. Ready-Made Culture Media

Media used in this study listed in (3.1.3) were prepared in accordance with the manufacture's instruction fixed on their containers.

All the media were sterilized by autoclaving at $121C°$ for 15 minutes. After sterilization, blood agar base was supplemented with 5% sheep blood, and urea agar base was supplemented with 2% sterile urea solution.

3.2.3.2. Laboratory –Prepared Culture Media

3.2.3.2.1 Motility Medium

Agar-agar (0.5 gm) was dissolved in 100 ml of brain heart infusion broth, the contents were dispensed into test tubes and autoclaved at 121 $C°$ for 15 minutes .This medium was used to detect the motility of bacteria (MacFaddin, 2000).

3.2.3.2.2. Gelatin Agar Medium

It was prepared by adding 4.4% gelatin to the nutrient agar medium, autoclaved and poured into sterile Petri dishes. It was used for detecting the bacterial ability to produce gelatinase enzyme (Collee *et al.*, 1996).

3.2.3.2.3. Esculin Agar Medium

This medium was prepared by dissolving 0.5 gm of esculin, 0.5 gm $FeCl_3$ and 40 gm of brain –heart infusion agar powder in 1000 ml of D.W. The contents were dispensed into test tubes, autoclaved and then prepared as slants.This medium was used for detecting the bacterial ability to hydrolyse esculin (Benson,1998).

3.2.3.2.4. Phenylalanine Medium (MacFaddin,2000)

The base medium was prepared by dissolving the following ingredients in 1000 ml D.W.

DL- phenylalanine	2gm
Yeast extract	3gm
NaCl	5 gm
Na_2HPO_4	1 gm
Agar-agar	12 gm

The ingredients were dispensed (4 ml) into test tubes, then autoclaved at 121C° for 15 min, and solidified in slanted position. It was used for detecting the bacterial ability to produce phenylalanine deaminase.

3.2.3.2.5. Carbohydrates Fermentation Medium

The medium was prepared according to MacFaddin (2000) as follows:

a: Basal Medium

Beef extract	1gm
Peptone	10gm
NaCl	5gm
Phenol red	0.018 gm
D.W.	1000 ml

The pH of mixture was adjusted to 7.4 and distributed into test tubes. Durham tube was inserted at the bottom of each test tube. The tubes were then autoclaved at 121 C° for 15 minutes, and cooled 1o 56 C° in the water bath.

b: Sugar Solutions

One % of the following sugars were used: glucose lactose, arabinose, D-manitol, sucrose,D-sorbitol, trehalose, raffinose, rhamminose, cellobiose, D-xylose and salicin (0.5%).

All sugar solutions were sterilized by millipore filters (0.22 μm). 0.1ml of each sterile sugar solution was added to each tube containing basal medium.

3.2.3.2.6. Maintenance Medium

This medium consisted of nutrient broth as a basal medium supplemented with 15% glycerol. After autoclaving at 121 C° for 15min, and cooling to 56 C° in water bath, 5ml aliquots were distributed in sterile tubes, and kept at 4 C^o until used. This medium was used to preserve the bacterial isolates at deep freeze for long term storage (Collee *et al.*,1996).

3.2.4 Patients and Sample Collection

During the period of five months from April to August 2011, a total of 801 clinical samples were collected from patients hospitalized / or attended to different hospitals in Hilla city / Babylon Province, included: Babylon Teaching Hospital for Maternity and Pediatric, AL- Hilla Teaching Hospital, Merjan Teaching Hospital and Chest Diseases Center. For each patient, clinical records were reviewed , which including: name, age , gender, address, hospitalization and antibiotic receiving. Of 801 patients , 328 male, and 473 female were included in this study.

Types and numbers of clinical samples are listed in Table (3-1)

Table (3-1): Types and numbers of clinical samples included in the present study.

Sample type	No.
Stool	141
Sputum	128
Vagina	116
Burn	153
Urine	97
Wound	60
Blood	58
Ear	30
Eye	8
Throat	10
Total	801

Stool and sputum specimens were collected in sterile containers In vaginal infection, the swabs were inserted into the posterior fornix, upper part of the vagina and rotated three before withdrawing them. A vaginal speculum was also used to provide a clear sight of the cervix and the swabs were rubbed in and around the introits of the cervix and withdrawn without contamination of the vaginal wall. In burns and wound, swabs were taken from the depth of burn or wound. In urinary tract infection ,specimens of urine were generally collected in plastic universal sterile containers from mid stream then centrifuged at 5000 rpm for 15 min. From blood ,2-5 ml withdrawn by disposable syringe under aseptic technique. The blood samples were immediately inoculated into sterilized blood culture bottle containing brian heart infusion broth. After that, blood culture bottle was incubated at $37C^o$ for 24-72 hrs. In ear infections, the swabs were inserted into the middle ear and rotated three before withdrawing it and withdrawn without contamination from the external ear. For eye and throat infections ,samples were taken by sterile swabs with transport medium. All samples must be transported to the laboratory without delay .

3.2.5. Isolation and Identification of Bacterial Isolates

All samples were streaked on MacConkey's agar and Blood agar,then incubated at $37C^o$ under aerobic conditions for 24 hrs. Culture results were interpreted as being lactose fermenting and non- fermenting bacteria. Lactose – fermenting isolates were subcultured, incubated for additional overnights. All isolates were examined for colonial morphology, then microscopically by Gram's staining. Suspected bacterial isolates were identified to the level of species or subspecies by using the following biochemical tests according to Holt *et*

54

al.(1994),Baron and Finegold (1994),Collee *et al.* (1996) and MacFaddin (2000) ,confirmatory identification was carried out by VITEK 2 system following manufacturers instructions.

3.2.5.1. Biochemical Tests

3.2.5.1.1. Indole Production Test

Peptone water was inoculated with a single agar bacterial culture and incubated at 37 C° for 24-48 hrs. A few drops of Kovac's reagent were added to each tube. Formation of pink ring indicated a positive result (MacFaddin,2000) .

3.2.5.1.2. Methly Red Test

Methyl Red- Voges Proskauer broth was inoculated with a single agar culture and incubated at 37 C° for 48-72 hrs, five drops of methyl red solution were added to each tube, mixed, and the result was read immediately. Changing the color to red is indicating a positive result (MacFaddin,2000).

3.2.5.1.3. Voges-Proskauer Test

Methyl Red-Voges Proskauer broth was inocultated with a single agar culture and incubated at 37 C° for 48-72 hrs. One ml of 40% KOH solution and 3ml of 5% α -naphthol solution were added to each tube. A positive reaction was indicated by the development of a dark red color in 20 minutes (MacFaddin, 2000).

3.2.5.1.4. Simmon's Citrate Test

Simmon's citrate slant was inoculated with a pure isolated colony and incubated at 37 C° for 48-72hrs. Changing the color of the medium from green to blue indicates a positive result (MacFaddin, 2000).

3.2.5.1.5. Kligler's Iron Agar Test

With straight inoculation needle, an inoculum was stabbed into the butt of the tube and streaked over the surface of the slant. Tubes were incubated at 37 C° for 24 hrs. Results were uniformatted according to MacFaddin (2000) as follows:

Slant / Butt	Color
Alkaline/ Acid	Red / Yellow
Acid / Acid	Yellow / Yellow
Alkaline / Alkaline	Red/ Red
H_2S production	Black precipitate
Gas production	Gas bubbles in the buttom

3.2.5.1.6. Urease Test

The surface of urea slant was streaked with a single bacterial culture and incubated at 37 C°. The result was read after 6 hrs , 24 hrs, and every day for 6 days. Changing the color of medium to purple-pink indicate a positive result (MacFaddin, 2000).

3.2.5.1.7. Malonate Utilization Test

Malonate broth was inoculated with a single bacterial culture and incubated at 37 C° for 48 hrs. Changing the color of medium from green to blue indicates a positive result (Collee *et al.*, 1996).

3.2.5.1.8. Motility Test

Tubes of Motility medium (3.2.3.2.1.) were inoculated with a single bacterial culture by stabbing vertically into the center of the medium and incubated at 37 C° for 24 – 48 hrs. Cloudy growth formation out of the line of stab indicates a positive result (MacFaddin,2000).

3.2.5.1.9. Gelatin Liquefaction Test

Gelatin agar (3.2.3.2.2.) was streaked with bacterial culture and incubated at 37 C° for 3-5 days. The plate was flooded with gelatin liquefaction reagent (3.2.2.5.) for 5-10 min. Appearance of clear zone around the growing colonies indicates production of gelatinase enzyme (Collee *et al.*, 1996).

3.2.5.1.10. Phenylalanine Deaminase

Phenylalanine medium (3.2.3.2.5.) was inoculated with a single bacterial culture by a straight wire, incubated at 37 C^o for 24 hrs, then the reagent was added directly over the slant. Changing the color of the slant to deep green indicates a positive result.

3.2.5.1.11. Carbohydrate Fermentation Test

Tubes of carbohydrate fermentation broth were inoculated with a single bacterial culture and incubated at 37 C^o for 1-5 days. Changing color of the indicator to yellow with or without gas production indicate a positive result (MacFaddin, 2000).

3.2.5.1.12. Esculin Hydrolysis Test

Esculin medium (3.2.3.2.3.) was inoculated with a single bacterial culture and incubated at 37 C^o for 24 hrs. Changing the color of medium and colonies to black indicates esculin hydrolysis.

3.2.5.1.13. Ornithine Decarboxylase Test

Ornithine decarboxylase medium was inoculated with a pure isolated colony, incubated at 37 C^o and read daily for 4 days. The color change of the medium from yellow to violet indicates a positive result (Collee *et al.*, 1986).

3.2.5.1.14. Growth at 10 C^o

Bacterial culture was streaked on nutrient agar plate and incubated at 10 C^o for 24 hrs. Appearance of bacterial colonies indicates the ability of isolate to grow in this low temperature (MacFaddin, 2000).

3.2.6. Preservation and Maintenance of Bacterial Isolates

The bacterial isolates were preserved on nutrient agar slant at 4 C^o. The isolates were maintained monthly by reculturing on fresh medium. Nutrient broth supplemented with 15% glycerol was used for long preservation and the isolates were maintained frozen at -20 C^o (deep freeze) for several months (long term maintenance) (Collee *et al.*, 1996).

3.2.7. Subculture of Frozen Stock Cultures

Frozen stock cultures were sub-cultured on fresh blood agar plates, then incubated in aerobic condition at 37 C° for 24hrs (Thomas, 2007).

3.2.8. Screening Test for β -lactam Resistance

Ampicillin and amoxicillin were added separately , from stock solution to the cooled Mueller- Hinton agar at a final concentration of 100 and 50 μg /ml, respectively. The media poured into sterilized Petri dishes , then stored at 4 C°. Preliminary screening of *K.pnenmoniae* isolates being resistant to β -lactam antibiotic was carried out using pick and patch method on the above plates (NCCLs,2003). Results were compared with *E.coli* ATCC 25922 as a negative control.

3.2.9. β -Lactamase Production

All bacterial isolates that were resistant to β - lactam antibiotic (3.2.8.) were tested for β - lactamase production by rapid iodometric method as follows:

Several colonies of a young bacterial culture on MacConkey's agar were transferred to Eppendrof's tube containing 100 μl of penicillin G solution , tubes were incubated at 37 C° for 30 minutes. Then, 50 μl of starch solution was added and mixed well with the content of the tube, 20 μl of iodine solution was added to the tube which causes the appearance of dark blue color , rapid change of this color to white (within few second to 2 minutes) indicated a positive result (Collee *et al*.,1996).

3.2.10. Antibiotic Susceptibility Testing

Antimicrobial susceptibility testing of β -lactam resistant *K.pneumoniae* isolates was performed on Mueller-Hinton agar plates by using (Kirby-Bauer) disk diffusion method against antibiotic listed in

58

(3.1.4.1). The cultures were incubated at 37 C° for 18 hrs under aerobic conditions and bacterial growth inhibition zones diameter were measured and interpreted in accordance with the Clinical and laboratory Standards Institute (CLSI) guidelines (CLSI, 2010). *E. coli* ATCC 25922 was used as the reference strain for antibiotic susceptibility testing.

3.2.11. Extended – Spectrum β - Lactamase Production

3.2.11.1. Initial Screening for ESBL Production

All bacterial isolates that resist β - lactam antibiotics were tested for ESBL production by initial screen test. The isolate would be considered potential ESBL producer, if the inhibition zone of ceftazidime (30 µg) disks was ≤22mm (CLSI, 2010).

3.2.11.2. Confirmatory Test

All bacterial isolates that resist β - lactam antibiotics were tested also for confirmatory ESBL production by two methods, these tests included:

a) Detection of ESBL by CHROMagar Technique

Freshly prepared extended spectrum β - lactamase CHROMagar plates were streaked by overnight growth of *K.pneumoniae*. The plates were incubated at 37 C° for 24 hrs according to manufacturer procedure. Growth of blue colonies indicated to ESBL producer. The reference strain of *E. coli* ATCC 25922 was inhibited and used as negative control.

b) Disk Approximation Test

This test was carried out as modified by Coudron *et al.* (1997). Antibiotic disks containing 30 µg cefotaxime, ceftazidime, ceftriaxone and azetreonam were placed 15mm (edge to edge) around a central disk of amoxi- clav (20 µg amoxicillin plus 10 µg clavulanate) on Mueller- Hinton agar plates inoculated with organism being tested for ESBL production.The plates were incubated aerobically at 37 C° for 24 hrs. Any augmentation (increase in diameter of inhibition zone) between the central amoxi-clav disk and any of the β - lactam antibiotic disks showing resistance or intermediate susceptibility was recorded, and the organism was thus considered as an ESBL producer.

3.2.12. Detection of Carbapenemases

a) Imipenem – EDTA Double Disk Synergy Test

Metallo β -lactamases detection was performed by double disk synergy method according to Lee *et al.*(2003). A 10 μg imipenem disk was placed in the center of a Mueller- Hinton agar plate inoculated with a 0.5 McFarland's tube dilution of the test isolate. An EDTA disk (1,900 μg) was placed at a distance of 15mm center to center from the imipenem disk. The plate was incubated at 37 C° overnight. The zone around the imipenem disk would be extended on the side is a metallo- β -lactamase producer.

b) Modified Hodge's Test (MHT)

This test was performed as described by Lee *et al.*, (2001). A 0.5 McFarland's tube dilution (3.2.1.1) of *E.coli* ATCC 25922 was prepared in 5 ml of tryptic soy broth, a lawn was streaked to a Mueller-Hinton agar plate and allow to dry (3-5) min, 10 μg imipenem disk was placed in the center of the test area. In a straight line the test organism streaked from the edge of the disk to the edge of the plate, the plates were incubated at 37C° overnight. Up to four organisms can be tested on the same plate with one disk. MHT positive test has a clover leaf-like indentation of the *E.coli* ATCC 25922 growing along the test organism growth streak within the disk inhibition zone. MHT negative test has no growth of the *E. coli* ATCC 25922 along the test organism.

c)Detection of KPC by CHROMagar Technique

K.pneumoniae carbapenemase (KPC) CHROMagar plates were streaked in the same day of preparation by overnight growth of *K.pneumoniae* , and incubated at 37 C° for 24 hrs according to manufacturer procedure. Growth of blue colonies indicated to suspected KPC producer. The reference strain of *E.coli* ATCC 25922 was inhibited and used as negative control.

3.2.13. Detection of AmpC β -Lactamase

3.2.13.1. Initial Screening AmpC β- Lactamase (Cefoxitin Susceptibility)

β - lactam resistant *K.pneumoniae* isolates were tested for cefoxitin susceptibity using standard disk diffusion method (CLSI, 2010). Isolates showing resistance to cefoxitin (inhibition zone diameter <18 mm) were considered as initially AmpC β - lactamase producers (Coudron *et al.*, 2003).

3.2.13.2. Confirmatory Tests of AmpC β - lactamase

β -lactam resistant isolates were tested for AmpC β -lactamase production by three confirmatory methods. These tests included:

a) Modified Three- Dimensional Test (MTDT)

This test was carried out according to Manchanda and Singh (2003).

Fresh overnight growth from Mueller-Hinton agar plate was transferred to a pre- weighed sterile Eppendrof's tube. The tube was weighed again to ascertain the weight of the bacterial mass. The technique was standardized so as to obtain 15 mg of bacterial wet weight for each sample. The growth was suspended in peptone water and pelleted by centrifugation at 3000 rpm for 15 minutes, bacterial growth washed with normal saline 2 to 3 times, crude enzyme extract was prepared by repeated freez thawing (approximately 10 cycles). Lawn cultures of *E.coli* ATCC 25922 were prepared on Mueller-Hinton agar plates and cefoxitin (30 µg) disks were placed on the plate. Linear slits (3 cm) were cut using a sterile surgical blade 3 mm away from the cefoxitin disk. Small circular wells were made on the slits at 5mm distance, inside the outer edge of the slit by stabbing with a sterile Pasteur pipette on the agar surface. Approximately 30 µl of extract was loaded in the wells, the plates were kept upright for (5-10) minutes until the solution dried, and then incubated at 37C° overnight. The isolates showing clear distortion of the zone of inhibition of cefoxitin were taken as AmpC producers. The isolates with no distortion were taken as AmpC non producers and isolates showing minimal distortion were taken as indeterminate strains.

b) AmpC Disk Test

All isolates subjected to MTDT were also simultaneously checked by AmpC disk test. A lawn culture of *E.coli* ATCC 25922 was prepared on Mueller-Hinton agar plate. Sterile disks (6 mm) were moistened with sterile saline (20 µl) and inoculated with several colonies of each test organism. The inoculated disk was then placed beside a cefoxitin disk (almost touching) on the inoculated plate. The plates were incubated overnight at 37 C° . A positive test appeared as flattening or indentation of the cefoxitin inhibition zone in the vicinity of the test disk. A negative test had an undistorted zone (Parveen *et al.*, 2010a)

c) Ceftazidime- Imipenem Antagonism Test (CIAT)

This test was carried out for detection of inducilde AmpC β - lactamases according to Cantarelli *et al* (2007) as follows:

An imipenem disk (10 µg) was placed 20 mm apart (edge- to –edge) from a ceftazidime disk (30 µg) on a Mueller-Hinton agar plate previously inoculated with a 0.5 McFarland's bacterial suspension, and incubated for 24 hrs at 35 C°. For comparison a cefoxitin disk was also placed 20 mm apart from the ceftazidime disk. Antagonism indicated by a visible reduction in the inhibition zone around the ceftazidime disk adjacent to the imipenen or cefoxitin disks, was regarded as positive for the inducible AmpC β -lactamases production.

3.2.14. DNA Extraction

The DNA extraction from bacterial cells was performed by two methods and used as a template for PCR amplification.

a) Genomic DNA Extraction Kit

DNA was purified by using the genomic DNA extraction kit according to manufacturer's instructions as follows:

- Cultured bacterial cells (up to 1×10^9) was transferred to a 1.5 ml microcentrifuge tube.
- The tubes were centrifuged for 1 minute at 14-16.000×g and the supernatant was discarded.
- Two hundred μl of GT Buffer was added to the tubes and the cell pellet was suspended by shaking vigorously or pipetting.
- The tubes were incubated at room temperature for 5 minutes.
- Two hundred μl of GB Buffer was added to the sample and mixed by shaking vigorously for 5 seconds.
- The tubes were incubated at 70 C° for 10 minutes or until the sample lysate is clear. During incubation , the tube was inverted every 3 minutes. At this time, the Elution Buffer (200 μl per sample) was incubated at 70 C° for DNA Elution.
- After 70 C° incubation, 5 μl of RNase A (10 mg /ml) was added to the sample lysate and mixed by vortex.
- The tubes were incubated at room temperature for 5 minutes.
- Two hundred μl of absolute ethanol was added to the sample lysate and immediately mixed by shaking vigorously.

- A GD column was placed in a 2 ml collection tube.
- All of the mixture (including any precipitate) was transferred to the GD Column.
- The tubes were centrifuged at 14-16,000×g for 2minutes.
- The 2ml collection tube containing the flow- through was discarded and the GD column was placed in a new 2 ml collection tube.
- Four hundred µl of W1 Buffer was added to GD column.
- The tubes were centrifuged at 14-16.000× g for 30 seconds.
- The folw- through was discarded and the GD column back was placed in the 2 ml collection tube.
- Six hundred µl of Wash Buffer (ethanol added) was added to the GD column.
- The tubes were centrifuged at 14-16.000×g for 30 seconds.
- The flow – through was discarded and the GD column back was placed in the 2 ml collection tube.
- The tubes were centrifuged again for 3 minutes at 14-16.000×g to dry the column matrix.
- The dried GD column was transferred to a clean 1.5 ml microcentrifuge tube.
- One hundred µl of preheated Elution Buffer or TE was added to the center of the column matrix.
- The tubes were left for 3.5 minutes or until the Elution Buffer or TE is absorbed by the matrix.
- The tubes were centrifuged at 14-16.000×g for 30 seconds to elute the purified DNA.

b)Salting Out Method

This method was performed according to Pospiech and Neuman (1995) with some modifications as follows:

A loopfull of *K. pneumoniae* overnight growth were inoculated in 5 ml nutrient broth and incubated at 37 C° for 24 hrs. The bacterial growth was centrifuged at 6000 rpm for 5 min. The precipitate was washed twice in 5 ml of TE buffer, then the pellet was resuspended in 2.5 ml SET buffer. 300 μl of freshly made (25%) SDS was added, mixed by inversion to the cell suspension, and incubated for 5 minutes at 55 C°. Then 1ml of 5M NaCl solution was added to the lysate, mixed thoroughly by inversion and let to be cooled to 37 C°. A mixture of equal volume (3.8 ml) of [phenol: chloroform: isoamylalcohol (25:24:1)] was added to the lysate and mixed by inversion for 30 min at room temperature. It was spun by centrifuge at 6000 rpm for 15 minutes. The aqueous phase was transferred to a fresh tube, which contains nucleic acid. Isoproponal (0.6 volume) was added to extract and mixed by inversion for 3 minutes; the DNA spooled on to a sealed pasture pipette. DNA rinsed in 5 ml of 70% ethanol, air dried for 15 min, and dissolved in 500 μl TE buffer at 4 C° overnight, then the DNA extract was stored in freezer at -20 C° until used.

3.2.15.Meaturement of DNA Concentration

DNA quantity and integrity was determined using a spectrophotometer (Nano drop) as the following:

1- Add 1 μl of TE solution on the lens for empty apparatus, be careful don't touch the lens.

2- 1 μl of DNA sample were added after that on the lens and read the absorbance at 260/280 nm.

3- Record the concentration of DNA at ng/ml.

3.2.16. Polymerase Chain Reaction Assay

3.2.16.1. Primers preparation

The Bioneer DNA primers were prepared depending on manufacturer instruction by dissolving the lyophilized product with D.D.W after spinning down briefly. Working primer tube was prepared by diluting with D.D.W. The final picomoles depended on the procedure of each primer (Appendix 2).

3.2.16.2. PCR Supplies Assembling and Thermocycling Conditions.

The DNA extracted from bacterial isolates were subjected to PCR using 14 sets (F and R) of primers for detection of β - lactam antibiotics resistance genes listed in (3.1.7.3). The reaction mixture moreover contain Go Taq® Green Master Mix, 2X which is premixed ready- to – use solution containing bacteriology derived *Taq* DNA polymerase dNTP, MgCl₂, and reaction buffers at optimal concentrations and its recommended for any amplification reaction to visualized by agarose gel electrophoresis and ethidium bromide staining.

Assembling PCR materials were done according to the procedure of Promega corporation (USA), using PCR reaction mixtures prepared in 0.2 ml Eppendrof tube with 25 µl reaction volumes which contain :12.5 µl Go Taq® Green Master Mix, 2X ,2.5 µl upstream primer,2.5 downstream primer,5 µl DNA template and 2.5 µl nuclease-free water (Appendix 2). All the appending was done in laminar flow on ice.

3.2.16.3. PCR Cycling Profiles

Polymerase chain reaction assays were carried out in a 25 µl reaction volume, and the PCR amplification conditions performed with a thermal cycler were specific to each single primer set depending on their reference procedure as in Table (3-3).

3.2.16.4. Agarose Gel Perparation

Agarose gel was prepared by dissolving 1.5 gm of agarose powder in 100 ml of TBE buffer previously prepared (90 ml D.W. were added to 10 ml TBE buffer 10X,the final concentration was 1X and pH =8.0). The mixture was placed in boiling water bath until it becomes clear, then allowed to cool to 50 C^o and 5 µl ethidium bromide at concentration of 0.5 mg/ml was added.

The agarose poured kindly in equilibrated gel tray earlier set with comb fixed in the end of the tray, and the two ends of gel tray were sealed. The agarose allowed to solidify at room temperature for 30 min. The comb and the seal were removed gently from the tray. The comb made wells used for loading DNA samples. The gel tray was fixed in an electrophoresis chamber which was filled with TBE buffer covering the surface of the gel.

3.2.17. PCR Product Analysis

3.2.17.1.Agarose Gel Electrophoresis

The amplified PCR products were detected by agarose gel electrophoresis and visualized by staining with ethidium bromide. PCR products were loaded to the agarose gel wells: 5 µl from single product to single well in known sequence, followed by 100 bp ladder to one of wells. The electric current was performed at 70 volt for 2-3hrs (Sambrook and Rusell, 2001).

3.2.17.2. Electrophoresis Results

The electrophoresis result was detected by using gel documentation system. The base pair of DNA bands were measured according to the ladder. The positive results were distinguished when there was DNA band equal to the target product size. Finally, the gel was photographed using gel documentation saving picture.

Table (3-2): PCR thermocycling conditions for detection *bla* genes

bla gene	Temperature (C ^{0}/Time)					Cycle number
	Initial denaturation	Cycling condition			Final extension	
		Denaturation	Anneling	Extension		
KPC	94/3min	94/1min	52/1min	72/1min	72/10min	30
SME	95/10min	94/30sec	58/30sec	72/30sec	72/10min	30
IMP	95/5min	95/20sec	59/40sec	72/30sec	72/5min	35
VIM	55/45sec	55/45sec	55/45sec	72/1min	72/10min	35
NDM-1	94/1min	94/1min	55/1min	72/2min	72/10min	30
OXA-23	95/30sec	95/30sec	51/1min	72/1min	72/10min	30
TEM	94/30sec	94/30sec	45/1min	72/1min	72/10min	35
SHV	94/30sec	94/30sec	60/1min	72/1min	72/10min	35
CTX-M	94/30sec	94/30sec	60/1min	72/1min	72/10min	35
VEB	93/3min	93/min	55/1min	72/1min	72/7min	40
PER	93/3min	93/1min	55/1min	72/1min	72/7min	40
GES	95/5min	94/1min	55/1min	72/1.5min	72/10min	40
OXA-1	94/5min	94/50sec	55/50sec	72/1min	72/10min	30
AmpC	94/30sec	94/30sec	60/1min	72/1min	72/10min	35

3.2.18.Determination of MICs of Carbapenemase Positive *K.*

pneumoniae Isolates.

3.2.18.1.Preparation of Bacterial Inoculum

Bacterial inoculum was prepared by transfer 4-5 similar colonies to 5 ml of tryptone soya broth and incubate at $35C^o$ for 2-8 hrs, the density of the suspension was adjusted depending on 0.5 MacFarland's standards by adding sterile normal saline.

3.2.18.2.HiComb MIC Test

HiComb consists of a strip made of an inert material , with 8 extensions that carry the discs of 4 mm, resembling the tooth of a comb. Towards the stem of the strip, MIC reading scale in µg/ml is given along with the Himedia code. A defined concentration of antibiotic is loaded on each of the disc so as to form a gradient when placed on agar plate. HiComb (based on the principle of dilution and diffusion) consists of a gradient that covers a continuous range of 16 two- fold dilutions on 2 different strips (Part A and B) as per the conventional MIC method. When applied to the agar surface , the antibiotic instantaneously diffuses into the surrounding medium in high to low amounts from one end of the strip to the other. The gradient remains stable after diffusion , and the zone of inhibition created takes the form of ellipse. This test is rapid than any agar or broth dilution method, and has a special advantage to study the resistance surveillance. The wide concentration gradients of these tests cover the MIC ranges of susceptibility of a wide variety of pathogens and allow both low level and high- level resistance detection (Himedia manufacture).

3.2.18.3.Using HiComb Strips

- The pack of HiComb was opened under aseptic conditions .
- One strip was picked up by a sterile forceps and placed on agar medium with its higher concentration facing the edge of plate and the markings on strip facing up wards.
- The strip was pressed gently on the handle and assured that all the discs were in full contact with the medium.
- Other strip was placed on the opposite side of plate with higher concentration towards the edge of plate and lower concentration towards the centre.
- The plate was closed and inverted to check whether all the disc are in full contact with the medium, 6 strips (3 of A and 3 of B) can be placed on a 21 cm plate with the markings facing upward.

For imipenem and meropenem Hicomb MIC strips were unavailable, instead Minimum Inhibitory Concentration Evaluator was used.

3.2.18.4. Minimum Inhibitory Concentration Evaluator

Minimum Inhibitory Concentration Evaluator (M.I.C.E) is a system for quantitatively determining the Minimum Inhibitory Concentration (MIC) of an antibiotic against a test organism.

3.2.18.4.1.Principle of The Test

M.I.C.E provides a gradient of antibiotic stabilized on a plastic strip with 30 graduations , to give an accurate MIC over the range 256 μg/ml-0.015 μg/ml. On application of the M.I.C.E strip to pre- inoculated agar , the antibiotic starts to release from the plastic forming a defined concentration gradient in the area around the strip. After appropriate incubation of the test, a zone of inhibition will have formed with the

M.I.C.E at the centre. The MIC can be easily read, using the graduated scale, at the point where the growth of the test organism touches the strip.

3.2.18.4.2.Inoculum

McFarland's inoculum level (0.5) was prepared . A sterile cotton swab was immersed in bacterial suspension and the excess moisture was removed by pressing against the edge of the tube. The plate was inoculated by swabbing in at least three different directions. The surface of the agar was left to dry completely before applying the M.I.C.E since excess moisture can cause a distortion of the gradient.

3.2.18.4.3.Application of The M.I.C.E

The strip should be applied to the plates within 15 min. of incubation to avoid pre- growth of the organism. Using sterile forceps, the M.I.C.E. was removed from the Sachet by handling the end with the logo and antibiotic code. The strip with the scale facing upwards was placed on agar surface, and the antibiotic gradient downwards in contact with the agar. The strip was applied by putting the end with the lowest concentration onto the plate first and then carefully rolling the strip onto agar to ensure good contact with the entire length of the M.I.C.E. Once the strips are applied to the agar, the plates should be incubated immediately in an inverted position at $37C^{o}$ for 18 hrs to avoid pre-diffusion of the antibiotic.

3.2.19. Transfer of Carbapenem Resistance

Conjugation experiment was carried out by liquid mating assay as described by Sambrook *et al.*, (1989) and Sambrook and Rusell, (2001). Two carbapenemase-positive isolates were selected as donor cells and rifampicin resistant *E.coli* MM294 was used as the recipient strain , Luria-Bertani (LB) agar plates containing ampicillin 100 µg /ml, ceftazidime 4 µg/ml, rifampicin 100µg/ml, and ampicillin 50 µg/ml were used for selection as required.

3.2.19.1.Transconjugation Experiment

- The ability of the donor isolates to grow on 100 µg/ml rifampicin containing LB agar plates was examined.
- Brain heart infusion agar plates were inoculated from donors and the recipient separately and incubated overnight at 37 C° .
- Five ml of LB broth media were inoculated with 4-5 colonies from the agar plates and incubated over night at 37 C° in shaking incubator.
- The overnight cultures of donor and recipient were diluted 100 folds and incubated at 37C°, shaking until the cultures were in exponential growth (OD_{600}:0.3-0.5). This was obtained after approximately 2 hrs of incubation .
- The donor and the recipient were mixed in a ratio 1:9 (0.5 ml of the donor was added to 4.5 ml of the recipient) and the incubation was continued at 37 C°, without shaking.
- After 60 minutes , the tubes were shacked well by vortex to terminate conjugation and ten-fold serial dilution of 10^{-1} to 10^{-6} were spread on the agar surface of Luria- Bertani agar supplemented with ceftazidime (4 µg/ml), rifampicin (100 µg/ml) and ampicillin (50 µg/ml) for transconjugant.
- 0.1 ml of each original bacterial culture of donor and recipient were spread on the selective medium mentioned earlier for counting spontaneously mutated cells. All the above-mentioned culture media were incubated at 37 C° for 24hrs.
- Plates were counted and transconjugation frequencies were calculated by dividing the number of transconjugants by the number of donors or by the number of recipients.

- Transconjugants were inoculated on LB agar plates containing 0.06 µg/ml meropenem to screen for transconjugants that co-acquired meropenem resistant of ≥ 0.06 µg/ml.
- DNA was extracted from transconjugant as mentioned earlier in (3.2.14.b) and served as a template in PCR experiment. Detection of carbapenemase ,ESBL and AmpC β - lactamase genes in transconjugants was carried out by PCR amplification using same primers and isolated template following same procedures had already applied during the screening.
- M.I.C.E test was used for detection the MIC for transconjugants.
- Stock cultures for transconjugants were prepared.

3.2.20. Statistical Analysis

The Chi-square test was performed for assessment of the statistical significance of the data. *P* value of < 0.05 was considered to be significant (Paulson, 2008).

4.1.Isolation and Identification of *K.pneumoniae* Isolates

The bacterial isolates obtained as a pure and predominant growth from clinical samples were only considered for the present study. Based on morphological and cultural properties of *Klebsiella* spp.(eg. isolates that exhibit large shiny pink mucoid colonies on MacConkey's agar ,no haemolysis on blood agar and large pink mucoid colonies without metallic shine on eosin methylen blue agar),as well as biochemical characterization, results revealed that only 177/801 (22%) isolates were belonged to *Klebsiella* spp., when 103/282 (36.5%) isolates were obtained from Babylon Teaching Hospital for Maternity and Pediatric, 51/261 (19%) isolates from Al-Hilla Teaching Hospital,4/130 (3%) isolates from Merjan Teaching Hospital and 19/128 (14.8%) isolates from Chest Diseases Center (Table 4-1).

Results from table (4-1) showed that 117/801 (14.6%) isolates were specified as *K. pneumoniae*. Of these 65/282 (23%) isolates were recovered from Babylon Teaching Hospital for Maternity and Pediatric, 29 /261 (11.1%) isolates from Al-Hilla Teaching Hospital,4 /130 (3%) isolates from Merjan Teaching Hospital and 19/128 (14.8%) isolate from Chest Diseases Center. However, prevalence and distribution of *Klebsiella pneumoniae* varied among Hilla hospitals.

Present results also revealed that a total of 569/801(71%) isolates were identified as other bacterial spp. and 55/801 (7%) samples showed no bacterial growth. Identification of *Klebsiella pneumoniae* isolates to the level of subspecies was carried out by the standard biochemical tests according to the Bergey's Manual of Determinative Bacteriology (Holt *et al.* ,1994) and MacFaddin (2000) (Appendix 1),then confirmed by VITEK 2 system.

Table (4-1): Distribution of bacterial isolates recovered from clinical samples among different hospitals in Hilla city.

Hospital's name	No. of sample	No. (%) of *Klebsiella* spp. isolates	No. (%) of *K. pneumoniae* isolates	No. (%) of other bacterial spp. isolates	No. (%) of no growth cultures
Babylon Teaching Hospital for Maternity and Pediatric	282	103 (36.5%)	65	179 (63.5%)	0 (0%)
Al- Hilla Teaching Hospital	261	51 (19 %)	29	195 (75%)	15 (6%)
Merjan Teaching Hospital	130	4 (3%)	4	86 (66%)	40 (31%)
Chest Diseases Center	128	19 (14.8%)	19	109 (85.2%)	0 (0%)
Total	801	177(22%)	117	569(71%)	55(7%)

Table (4-2) shows the distribution of *K.pneumoniae* isolates among different clinical samples, when 38(27%),19(15%),18(15.5%),18(11.7%) ,10(10%) 8 (13.3%),3(5%),2(6.6%) and 1(12.5%)were recovered from stool ,sputum, vagina , burn , urine, wound, blood, ear and eye samples , respectively.

Same table reveled that *K. pneumoniae* subsp. *pneumoniae* was the most prevalent subspecies of *K. pneumoniae* which comprised 77 (9.6%).

However , majority of isolates 26 (18 %) were obtained from stool samples, 8 (6%) from sputum , 18 (15.5 %) from vagina , 12 (7.8 %) from burn, 3 (3 %) from urine , 5 (8.3 %) from wound , 2(3%) from blood ,2 (6.6%) from ear and 1 (12.5 %) from eye.

Out of 117 *K. pneumoniae* isolates , 34 (4 %) were belonged to *K. pneumoniae* subsp *ozaenae* , of these 8 (6%) were isolated from stool , 9 (7 %) from sputum , 6 (3.9 %) from burn, 7 (7 %) from urine , 3 (5 %) from wound and 1 (2%) from blood ,whereas 6 (1%) isolates were identified as *K. pneumoniae* subsp . *rhinoscleromatis*, 4 (3 %) isolates from stool and 2 (2 %) isolates from sputum .

Table (4-2): Numbers and percentages of *K. pneumoniae* subspecies among different clinical samples.

Clinical sample	No. of sample	No. (%) of *K. pneumoniae* isolates	No. (%) of *K . pneumoniae* subspecies		
			K. pneumoniae subsp. *pneumoniae*	*K. pneumouiae* subsp. *ozaenae*	*K. pneumoniae* subsp. *rhinoscleromatis*
Stool	141	38 (27%)	26(18%)	8(6%)	4(3%)
Sputum	128	19 (15%)	8(6%)	9(7%)	2(2%)
Vagina	116	18 (15.5%)	18(15.5%)	0(0%)	0(0%)
Burn	153	18 (11.7%)	12(7.8%)	6(3.9%)	0(0%)
Urine	97	10 (10%)	3(3%)	7(7%)	0(0%)
Wound	60	8 (13.3%)	5(8.3%)	3(5%)	0(0%)
Blood	58	3 (5%)	2(3%)	1(2%)	0(0%)
Ear	30	2 (6.6%)	2(6.6%)	0 (0%)	0(0%)
Eye	8	1 (12.5%)	1(12.5%)	0(0%)	0(0%)
Throat	10	0(0%)	0(0%)	0(0%)	0(0%)
Total	801	117(14.6%)	77(9.6%)	34(4%)	6(1%)

4.2. Primary Screening Test of β - Lactam Resistant Isolates

All 117 *K. pneumoniae* obtained from different clinical samples were primarily screened for β - lactams resistance by growing on Mueller- Hinton agar supplemented with ampicillin and amoxicillin (each alone) at final concentrations of 100 and 50 μg /ml , respectively. Results from Table (4-3) revealed that a total of 91/117(78%) *K. pneumoniae* isolates were able to grow normally in the presence of ampicillin and amoxicillin.

Table (4-3): β - lactam resistant *Klebsiella pneumoniae* isolates recovered from different clinical samples.

No. of *K. pneumoniae* isolates	Susceptibility to ampicillin and amoxicillin	
	No. (%) of resistant isolates	No. (%) of sensitive isolates
117	91 (78%)	26 (22%)

4.3. Antibiotic Susceptibility Test of *K. pneumoniae* Isolates

As determined by disk-diffusion antibiotic susceptibility testing, all 91 β -lactam resistant *K. pneumoniae* isolates exhibited different pattern of resistance to β - lactam agents (Figure 4-1), demonstrating higher resistance to penicillins (carbenicillin and ampicillin) with rates of resistance of 90(99%) and 86(94.5%),respectively, whereas 75(82.4%) of isolates were resistance to piperacillin.

Resistance to other drug classes varied among the isolates. For cephalosporin generations (third and fourth) ,a higher resistance was also detected with 79(86.8%) of isolates being resistant to ceftazidime , 76(83.5%) to cefotaxime ,75(82.4%) to ceftriaxone and 73(80.2%) to cefepime. The results also revealed that were high resistant rates for amoxi-clav 74(81.3%) and cefoxitin 71(78%). A 72(79.1 %) and 72(79%) resistance were noticed to cefaclor ,cefprozil and aztreonam antibiotics, respectively.

Among the carbapenem ,imipenem displayed a lower resistance rate 9(10%),than meropenem 16(17.6%) and ertapenem 17(18.7%). Aminoglycosides resistance was variable ,46(50%) to kanamycin , 37(40.6%) to gentamicin and 26 (28.6 %) to amikacin.

The resistance to quinolones, nalidixic acid ,ciprofloxacin and levofloxacin was detected 39 (42.8%), 30(33%), 26(28.5%), respectively. Percentages of resistance of isolates to the remaining antibiotics were as follows : tetracycline 57(62.6%), doxycycline and nitrofurantoin 54(59.3%) each, trimethoprim-sulfamethoxazole 51(56%) and chloramphenicol 36(39.6%). Results revealed that all tested isolates were resistant to a minimum of 3 classes of antibiotics, hence these isolates were considered to be multidrug resistant.

4.4. Detection of β - Lactamase Production

All *K. pneumoniae* isolates that resist β - lactam antibiotics were tested for β - lactamase production by using rapid iodometric method. The results showed that 77/91(84.6 %) of tested isolates were positive for β - lactamase production (Table 4-4). However ,14/91 (15.4 %) of the isolates gave negative results.

Table (4-4): β - lactamases producing *Klebsiella pneumoniae* isolates by rapid iodometric method.

No. β - lactam resistant *K. pneumoniae* isolates	No. (%) of β - lactamase producers	No. (%) of non β - lactamase producers
91	77 (84.6%)	14 (15.4%)

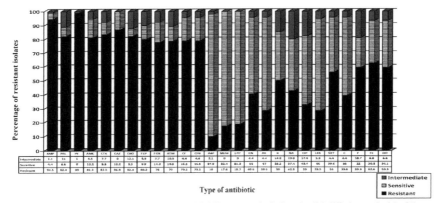

Figure (4-1): Antibiotics susceptibility profile of *Klebsiella pneumoniae* isolates by disk diffusion method (n=91).
AMP,Ampicillin;PRL,Piperacillin;PY,Carbenicillin;AMC,Amoxi-clav;CTX,Cefotaxime;CAZ,Ceftazidime;CRO,Ceftraiaxone;FEP
Cefepime; Fox,Cefoxitin; ATM,Aztreonam; CF,Cefaclor; CPR,Cefprozil; IMP,Imipenem; MEM,Meropenem; ETP,Ertapenem; CN,
Gantamicin ; AK, Amikacin ; K, Kanamycin ; NA , Nalidixic acid ; CIP, Ciprofloxacin ; LE⁵,Levofloxacin , SXT, Trimethoprim-
Sulfamethoxazole;C,Chloramphenicol;F,Nitrofurantion;TE, Tetracycline;DO,Doxycycline.

4.5. Extended – Spectrum β - Lactamases (ESBLs) Production

All 91 β - lactam resistant *K. pneumoiae* isolates collected from different hospitals in Hilla city were screened for ESBL production according to the Clinical and Laboratory Standard Institute criteria (CLSI,2010). ESBL production among these isolates were detected phenotypically by two methods, CHROMagar and disk approximation method.

In initial screening test, isolates of *K. pneumoniae* were tested for ESBL production by using ceftazidime disks. According to the CLSI (2010), the isolate is considered to be a potential ESBL producer , if the inhibition zone of ceftazidime disks (30 µg) was ≤ 22mm. The results obtained from this study revealed that all β - lactam resistant *K. pneumoniae* isolates were found to be ESBL- producers , initially.

A positive result from initial screening was followed by a phenotypic confirmatory tests using ESBL CHROMagar and disk approximation methods. In CHROMagar confirmatory test, results of the present study confirmed the presence of an ESBL in 81/91(89%) of β - lactam resistant isolates (Table 4-5). All these isolates showed overnight growth with blue colonies on the ESBL supplemented CHROMagar orientation medium (Figure 4-2), which was interpreted as a phenotypic evidence of ESBL production.

In the disk approximation method, any enhancement of the inhibition zone between a β - lactam disks (30 µg) (ceftazidime , ceftriaxone , cefotaxime and aztreonam) toward amoxicillin- clavulanate disk (20 µg /10 µg) gave an indication that the test strain is ESBL producer (Figure 4-3).

Table (4-5): Frequency of ESBL- producing *K. pneumoniae* isolates by two phenotypic confirmatory methods .

No. of β - lactam resistant *K. pneumoniae* isolates	No. (%) of isolates detected by	
	ESBL* CHROMagar technique	Disk approximation method
91	81 (89%)	45(49.5%)

*ESBL :Extended-spectrumβ   - lactamase
χ^2 is significant in($p < 0.05$)

Figure (4-2): ESBL production by *K. pneumoniae* K13 test isolate exhibit deep blue colonies on the ESBL supplemented CHROMagar medium after incubation at 37C° for 24 hrs.

The results from Table (4-5) showed that out of 91 β - lactam resistant *K. pneumoniae* isolates , only 45 (49.5%) were confirmed to be ESBL producer by this method . However, the detection of ESBL production by CHROMagar technique was more significant (P <0.05) than the disk approximation method.

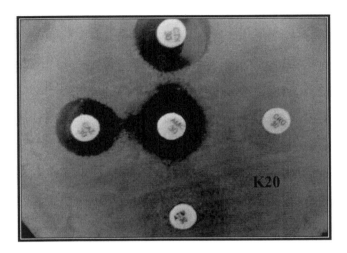

Figure (4-3): Disk approximation test for detection of ESBL in *K. pneumoniae* K 20 isolate .AMC, amoxi- clav; CAZ, ceftazidime;CTX, cefotaxime; CRO, ceftriaxone; ATM , aztreonam.

4.6. Carbapenemase Production

Among the 91 β - lactam resistant *K. pneumoniae* isolates, 17 had previously been identified as resistant to carbapenem antibiotics by standard disk diffusion method (Figure 4-1). Of these 9(10%),16(17.6%), 17(18.7%) isolates were resistant to imipenem ,meropenem and ertapenem ,respectively, and 9(10%) were resistant to both imipenem and meropenem (Table 4-6).All these isolates were subjected to three phenotypic confirmatory screening tests for carbapenemase production, imipenem- EDTA double disk synergy test , modified Hodge's test and KPC CHROMagar.

Metallo β - lactamases detection was performed by using the imipenem- EDTA disk test. Results revealed that out of 17 carbapenem – resistant isolates, 11 (65%) showed an increase in the zone diameter around the IMP – EDTA disk, suggesting the production of metallo β - lactamases (Table 4-7and Figure 4-4).

In the modified Hodge's test, 14 (82%) isolates displayed a clover leaf like indentation with imipenem disk indicating a positive result (Table 4-7 and Figure 4-5).

Table (4-6):Carbapenem resistant profile of *Klebsiella pneumoniae* isolates

No.(%) of carbapenem - resistant isolates	No.(%) of isolates resistance to			
	imipenem	meropenem	ertapenem	imipenem and meropenem
17	9(10%)	16(17.6%)	17 (18.7%)	9(10%)

Table (4-7): Phenotypic characterization of carbapenem – resistant
***K. pneumoniae* by three confirmatory methods.**

No. of carbapenem- resistant isolates	No. (%) of carbapenemase producer isolates detected with		
	Imipenem- EDTA disk test	Modified Hodge's test	KPC[*] CHROMagar technique
17	11(65%)	14(82%)	17(100%)

*[*] KPC: *Klebsiella pneumoniae* carbapenemase*

The performance of the chromogenic medium , CHROMagar KPC was also evaluated for detection KPC- producing *K. pneumoniae*. All the screened isolates , 17 (100%) yielded overnight blue colonies on KPC supplemented CHROMagar medium (Table 4-7 and Figure 4-6).There is no significant difference ($p <0.05$) between three methods.

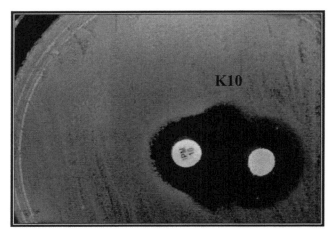

Figure (4-4): Phenotypic appearance of MBL – producing *K. pneumoniae* by imipenem – EDTA double disk synergy test. K10 test isolate showing large synergistic inhibition zone between IMP disk and EDTA disk (positive result).

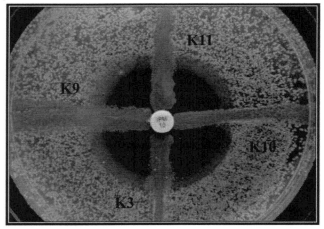

Figure (4-5): Phenotypic carbapenemase detection by modified Hodge's test. Growth of *E.coli* ATCC 25922 strain around straight line of test isolates , K10 and K11 isolates , showing clear distortion of the inhibition zone of imipenem (IMP) disk ,K3and K9 isolates showing negative result.

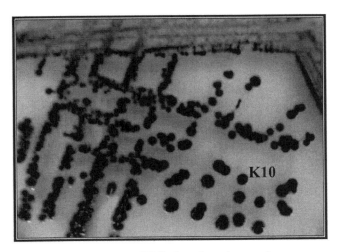

Figure (4-6): CHROMagar KPC for detection carbapenemase-producing *K. pneumoniae* isolate , K10 test isolate exhibit blue colonies indicated to KPC producer.

4.7.Molecular Detection of Carbapenemase Genes In Clinical Isolates of *K.pneumoniae*.

Detection of carbapenemase (KPC, VIM , IMP ,NDM-1, OXA-23 and SME) was performed by the conventional PCR technique. Table (4-8) shows the distribution of carbapenemase genes determind by the consistent results primers specific PCR.

The results revealed that all examined isolates of *K. pneumoniae* 17(100%) had carbapenemase genes with a varying presence. However,the gene that encodes VIM was identified among 14(82.3 %) isolates (Figure 4-7).Three (17.6%) of the examined isolates possessed *bla*NDM-1 (Figure 4-8),while the presence of the *bla*OXA-23 was detected in 15 (88.2%) isolates (Figure 4-9).

While, none of the 17 carbapenem-resistant *K.pneumoniae* isolates showed amplification for *bla*KPC, *bla*IMP and *bla*SME genes.

Table (4-8): Distribution of carbapenemase genes among carbapenem resistant *K. pneumoniae* isolates.

Isolate symbol	Types of carbapenemase genes					
	KPC	VIM	IMP	NDM-1	OXA-23	SME
K_1	-	-	-	-	+	-
K_2	-	-	-	-	+	-
K_3	-	+	-	-	+	-
K_4	-	+	-	-	+	-
K_5	-	+	-	-	+	-
K_6	-	+	-	-	-	-
K_7	-	+	-	-	+	-
K_8	-	+	-	-	+	-
K_9	-	+	-	-	-	-
K_{10}	-	+	-	-	+	-
K_{11}	-	+	-	-	+	-
K_{12}	-	+	-	-	+	-
K_{13}	-	+	-	+	+	-
K_{14}	-	+	-	+	+	-
K_{15}	-	-	-	+	+	-
K_{16}	-	+	-	-	+	-
K_{17}	-	+	-	-	+	-
Total positive	0 (0%)	14 (82.3%)	0 (0%)	3 (17.6%)	15 (88.2%)	0 (0%)

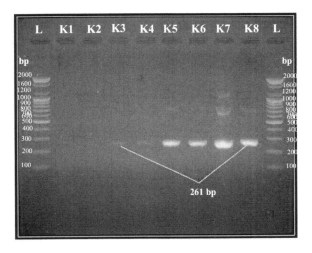

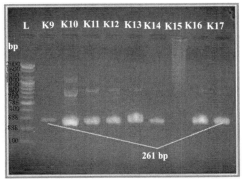

Figure (4-7): Agarose gel electrophoresis (1.5%agarose,70 % volt for 2-3 hrs) for *bla*VIM gene product (ampilified size 261 bp) using DNA template of carbapenem-resistant *K. pneumoniae* isolates extracted by using salting out method. Lane (L), DNA moleculer size marker (100- bp Ladder).Lanes (K3, 4, 5, 6, 7, 8, 9,10,11,12,13,14, ,16 and 17) of *K. pneumoniae* isolates show positive results with *bla*VIM gene , lanes (K1,2and 15) show negative results with *bla*VIM gene.

K1 K2 K3 K4 K5 K6 K7 K8 K9 K10 K11 K12 K13K14K15K16K17 L

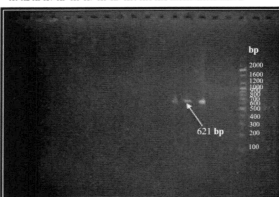

Figure (4-8): Agarose gel electrophoresis (1.5%agarose,70 % volt for 2-3 hrs) for *bla*NDM-1 gene product (ampilified size 621 bp) using DNA template of carbapenem-resistant *K. pneumoniae* isolates extracted by using salting out method. Lane (L), DNA moleculer size marker (100-bp Ladder).Lanes(K13,14 and 15) of *K. pneumoniae* isolates show positive results with *bla*NDM-1 gene , lanes (K1,2,3, 4, 5, 6, 7, 8, 9,10,11,12, 16 and 17) show negative results with *bla*NDM-1 gene.

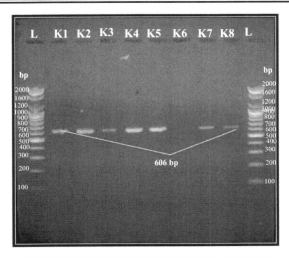

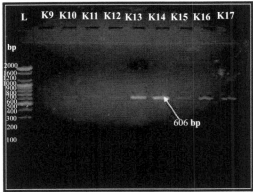

Figure (4-9): Agarose gel electrophoresis (1.5%agarose,70 % volt for 2-3 hrs) for *bla*OXA-23 gene product (ampilified size 606 bp) using DNA template of carbapenem-resistant *K. pneumoniae* isolates extracted by using salting out method. Lane (L), DNA moleculer size marker (100- bp Ladder).Lanes (K1,2,3, 4, 5, 7, 8, 10,11,12,13,14,15,16 and 17) of *K. pneumoniae* isolates show positive results with *bla*OXA-23 gene , lanes (K 6 and 9) show negative results with *bla*OXA-23 gene.

4.8.Emergence of *K.pneumoniae* Harboring Carbapenemase Genes

Table (4-9) showed that 17 carbapenemase positive *K.pneumoniae* isolates were obtained from 801 clinical samples. More isolates were detected among young and middle age groups,1-20 years 9(2.8%) and 21-40 years 5(1.9%).However, less isolates 3(1.6%) were detected in the age group of 41-60 years. Whereas, none of the isolates were obtained from patients with the age of 61-80 years 0(0%). It was found that patients with the age of 1-20 years were at increased risk for infection with carbapenemase harboring *K.pneumoniae* in comparison to other age groups.

Although, there was no indicative significance (p <0.05) among gender of patients, most of the patients from whom the isolates obtained were females 12 (2.5%) than males 5(1.5%). In term of hospitalization, 16 (2.6%) of 17 carbapenemase positive isolates were obtained from inpatients ,as compared with outpatients 1(0.6%).Moreover, all carbapenemase positive *K.pneumoniae* isolates were recovered from patients had previously taken different antibiotics 17 (2.1%).There is clear significant difference (p<0.05) was between existed carbapenemase positive isolates and age, hospitalization and antibiotic receiving.

Table (4-10) reveals the distribution of 17 carbapenemase positive *K.pneumoniae* isolates among different clinical samples .Results showed that 9/153 (5.9%) of them were isolated from burn ,3/60 (5%) from wound, 2/141(1.4%) from stool ,2/116 (1.7) from vagina and only 1/128 (0.8%) isolate from sputum samples (12 isolates were belonged to *K.pneumoniae* subspp.*pneumoniae* and 5 isolates were *K.pneumoniae* subspp.*ozaenae,* Appendix 3). Adversely, non of the carbapenemase positive *K.pneumoniae* isolates were detected among other samples collected from urine, blood, ear, eye and throat.

Table (4-9): Clinical profile of patients infected with carbapenemase genes positive *Klebsiella pneumoniae*

Patient profile	Status	No. of sample (n=801)	No. (%) of carbapenemase positive *K. pneumoniae* isolates (n=17)
Age group [*] (years)	1-20 21-40 41-60 61-80	316 261 192 32	9 (2.8%) 5 (1.9%) 3 (1.6%) 0 (0%)
Gender [**]	Male Female	328 473	5(1.5%) 12 (2.5%)
Hospitalization [*]	Inpatient Outpatient	625 176	16 (2.6%) 1 (0.6%)
Antibiotic [*] Administration	+ -	801 0	17 (2.1%) 0 (0%)

[*] χ^2 is significant in ($p < 0.05$)

[**] χ^2 is not significant in ($p < 0.05$)

Table (4-10):Distribution of carbapenemase positive *K.pneumoniae* isolates among different clinical samples

Sample source	No. of sample	No.(%) of carbapenemase positive *K.pneumoniae* (n=17)
Stool	141	2(1.4 %)
Sputum	128	1(0.8%)
Vagina	116	2(1.7%)
Burn	153	9 (5.9%)
Urine	97	0(0%)
Wound	60	3(5%)
Blood	58	0(0%)
Ear	30	0(0%)
Eye	8	0(0%)
Throat	10	0(0%)
Total	801	17 (2.1%)

4.9.Antibiotic Susceptibility of Carbapenemase Positive *Klebsiella pneumoniae* Isolates.

All 17 carbapenemase positive *K. pneumoniae* isolates were tested for antibiotic susceptibility towards 26 antimicrobial agents.

Results from Figure (4-10) showed that all these isolates (100%) were resistant to ampicillin , piperacillin, carbenicillin , amoxi – clav , cefoxitin, ceftazidime, cefotaxime , ceftriaxone, cefepime , cefaclor, cefprozil, aztreonam, ertapenem, tetracycline, doxycycline , trimethoprim-sulfamethoxazole and chloramphenicol. The next most resistant level were recorded for meropenem and kanamycin (94%each), gantamicin and nitrofurantion (88%, each) and ciprofloxacin (82%).

Resistance to amikacin, nalidixic acid and levofloxacin was shown in (70%) of these isolates . Moreover, (53%) resistance rate was observed for imipenem antibiotic.

4.10.Drug Resistance Pattern among Carbapenemase Genes Positive *K. pneumoniae* Isolates .

This study was sought to screen pattern of resistance to antibiotics among 17 carbapenemase genes positive *K. pneumoniae* isolates. A strain is considered a multidrug-resistant (MDR), if an isolate is resistant to representatives of three or more classes of antibiotics (penicillins, cephalosporins, aminoglycosides, monobactams , fluoroqinolones, tetracyclines and carbapenems) .

Extensive drug- resistant (XDR) was defined as resistant to at least one agent in all but one or two classes. A pandrug-resistant (PDR) isolate is resistant to all agents in all antibiotics classes (Falagas and Karageopoulos , 2008 and Magiorakos *et al.*, 2011).

All isolates were found to be resistant to at least 9 antibiotic classes tested .Among which 11 (65%) isolates were classified as XDR isolates. Of these 4(24%) isolates were resistant to 9 antibiotic classes, whereas 3(17%) and 4 (24%) were resistant to 10 and 11 antibiotic classes . Moreover , 6 (35%) of carbapenemase positive *K. pneumoniae* isolates were resistant to all available agents in all antibiotic classes tested and hence considered PDR isolates (Table 4-11).

Table (4-11) : Antibiotics resistant pattern of carbapenemase genes positive *K. pneumoniae* isolates.

Type of resistance	No. (%) of carbapenemase positive *K. pneumoniae* isolates (n=17)	No. of resisted antibiotic classes
* XDR	4(24%)	9
	3(17%)	10
	4(24%)	11
** PDR	6 (35 %)	12

* XDR: Extensive drug resistant; ** PDR: Pandrug resistant

4.11. Minimum Inhibitory Concentrations of Carbapenemase Positive *K.pneumoniae* Isolates

Minimum inhibitory concentrations of 17 carbapenemase positive isolates against selected beta-lactam antibiotics (imipenem, meropenem, ampicillin , cefotaxime,ceftriaxone ,and ceftazidime) were determined by using HiComb and Minimum inhibitory concenteration evaluator tests according to the Clinical and Laboratory Standards Institute (2010) guidelines. Table (4-12) revealed that all carbapenemase positive isolates were resistant to imipenem, meropenem, ampicillin cefotaxime, ceftriaxone and ceftazidime with concentrations reached beyond the break point values: >32 µg /ml ,>32 µg /ml,>256 µg /ml , >240 µg /ml ,>256 µg /ml and >240 µg /ml, respectively (Figures 4-11 and 4-12), except two isolates (K1 and K5) showed variation in the MIC values for cefotaxime,ceftriaxone and ceftazidime with 120 µg /ml,128 µg /ml and 60 µg /ml for each, respectively.

Table (4-12) : Minimum inhibitory concentrations of β –lactam antibiotics for carbapenemase positive *K. pneumoniae* isolates.

Isolate symbol	MIC (μg /ml)					
	IMP (≥16)	MEM (≥16)	AMP (>32)	CTX (>64)	CTR (>64)	CAZ (>32)
K2-4, K6-17	>32	>32	>256	>240	>256	>240
K1	>32	>32	>256	120	128	60
K5	>32	>32	>256	120	128	60

Number between brackets refer to break point values

4.12. Screening and Detection of ESBL Genes in Carbapenemases Harboring *K. pneumoniae* Isolates.

Presence of *bla*TEM, *bla*SHV, *bla*CTX-M, *bla*OXA-1, *bla*PER, *bla*VEB and *bla*GES were determined by PCR technique. Table (4-13) shows the distribution of ESBL genes among the isolates. However , results revealed that of 17 carbapenemase positive isolates analyzed, 13 (76.5 %) isolates yielded amplification products with TEM , SHV , CTX-M and OXA-1 PCR specific primers (Figures 4-13, 4-14, 4-15 and 4-16, respectively). Whereas, *bla*PER gene was observed in 10 (58.8%) isolates (Figure 4-17).

On the other hand , no ESBL genes , including *bla*VEB and *bla*GES were identified in the 17 carbapenemase positive *K. pneumoniae* isolates.

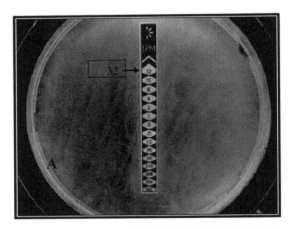

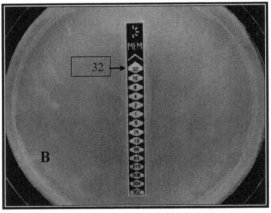

Figure (4-11): Minimum inhibitory concentrations by minimum inhibitory concenteration evaluator test for carbapenemase positive *K.pneumoniae* isolate. (A),MIC for IMP >32 μg /ml and (B), MIC for MEM >32 μg /ml. IMP, imipenem; MEM, meropenem .

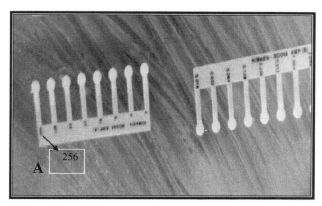

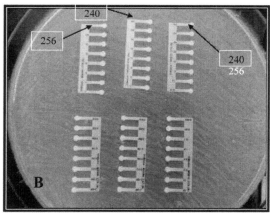

Figure (4-12): Minimum inhibitory concentrations by HiComb test for carbapenemase positive *K.pneumoniae* isolate.(A), MIC for AMP > 256 μg /ml (B),MIC for CTX >240 μg /ml,CAZ>240 μg /ml and CTR>256 μg /ml .AMP,ampcillin;CTX, cefotaxime; CAZ,ceftazidime; CTR, ceftriaxone.

Table (4-13) : Distribution of ESBL genes among carbapenemase positive *K. pneumoniae* isolates.

Isolate symbol	Types of ESBL genes						
	TEM	SHV	CTX-M	OXA-1	PER	VEB	GES
K1	-	-	-	-	-	-	-
K2	-	-	-	-	-	-	-
K3	+	+	+	+	-	-	-
K4	-	-	-	-	-	-	-
K5	+	+	+	+	-	-	-
K6	+	+	+	+	+	-	-
K7	+	+	+	+	+	-	-
K8	+	+	+	+	+	-	-
K9	+	+	+	+	+	-	-
K10	+	+	+	+	+	-	-
K11	+	+	+	+	+	-	-
K12	+	+	+	+	-	-	-
K13	+	+	+	+	+	-	-
K14	+	+	+	+	+	-	-
K15	-	-	-	-	-	-	-
K16	+	+	+	+	+	-	-
K17	+	+	+	+	+	-	-
Total positive	13 (76.5%)	13 (76.5%)	13 (76.5%)	13 (76.5%)	10 (58.8%)	0 (0%)	0 (0%)

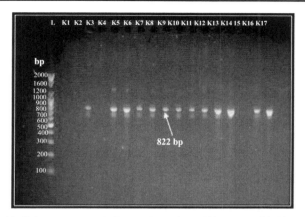

Figure (4-13): Agarose gel electrophoresis (1.5%agarose,70 % volt for 2-3 hrs) for *bla*TEM **gene product (ampilified size 822 bp) using DNA template of carbapenem-resistant** *K. pneumoniae* **isolates extracted by using salting out method . Lane (L), DNA molecular size marker (100-bp ladder). Lanes (K3, 5, 6, 7, 8, 9, 10, 11, 12, 13, 14, 16 and 17) of** *K. pneumoniae* **isolates show positive results with** *bla*TEM **gene. Lanes (K1, 2,4 and 15) show negative results with** *bla*TEM **gene.**

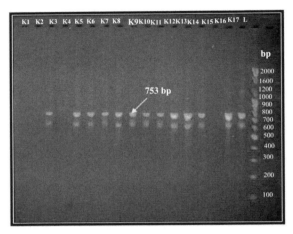

Figure (4-14): Agarose gel electrophoresis (1.5%agarose,70 % volt for 2-3 hrs) for *bla*SHV **gene product (ampilified size 753 bp) using DNA template of carbapenem-resistant** *K. pneumoniae* **isolates extracted by using salting out method. Lane (L), DNA molecular size marker (100-bp ladder). Lanes (K3, 5, 6, 7, 8, 9, 10, 11, 12, 13, 14, 16 and 17)of** *K. pneumoniae* **isolates show positive results with** *bla*SHV **gene. Lanes (K1, 2,4 and 15) show negative results with** *bla* SHV **gene.**

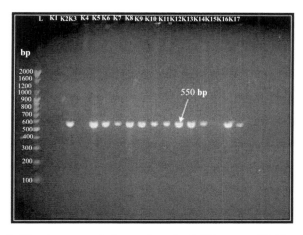

Figure (4-15): Agarose gel electrophoresis (1.5%agarose,70 % volt for 2-3 hrs) for *bla*CTX-M **gene product (ampilified size 550 bp) using DNA template of carbapenem-resistant** *K. pneumoniae* **isolates extracted by using salting out method. Lane (L), DNA molecular size marker (100-bp ladder). Lanes (K3, 5, 6, 7, 8, 9, 10, 11, 12, 13, 14, 16 and 17)of** *K. pneumoniae* **isolates show positive results with** *bla* CTX-M **gene. Lanes (K1, 2,4 and 15) show negative results with** *bla* CTX-M **gene.**

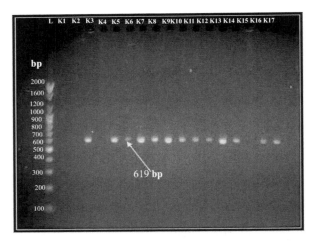

Figure (4-16) :Agarose gel electrophoresis (1.5%agarose,70 % volt for 2-3 hrs) for *bla*OXA-1 gene product (ampilified size 619 bp) using DNA template of carbapenem-resistant *K. pneumoniae* isolates extracted by using salting out method. Lane (L), DNA molecular size marker (100-bp ladder). Lanes (K3, 5, 6, 7, 8, 9, 10, 11, 12, 13, 14, 16 and 17)of *K. pneumoniae* isolates show positive results with *bla* OXA-1 gene. Lanes (K1, 2,4 and15) show negative results with *bla* OXA-1 gene.

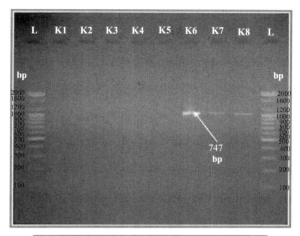

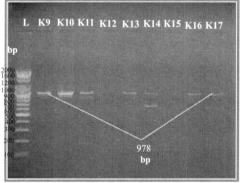

Figure (4-17): Agarose gel electrophoresis (1.5%agarose,70 % volt for 2-3 hrs) for *bla*PER gene product (ampilified size 978 bp) using DNA template of carbapenem-resistant *K. pneumoniae* isolates extracted by using salting out method. Lane (L), DNA molecular size marker (100-bp ladder). Lanes (K 6, 7, 8, 9, 10, 11, , 13, 14, 16 and 17)of *K. pneumoniae* isolates show positive results with *bla*PER gene. Lanes (K1, 2, 3, 4, 5, 12 and 15) show negative results with *bla*PER gene.

4.13. AmpC β -Lactamase Production among Carbapenemase

Positive *K. pneumoniae*

All carbapenemase positive isolates were tested for cefoxitin susceptibility by the standard Kirby- Bauer disk diffusion method. Isolates yielded cefoxitin zone diameter less than 18 mm, such isolates may be AmpC β - lactamase producers . Results showed that all carbapenemase positive *K. pneumoniae* isolates yielded cefoxitin zone diameter less than 18mm, these isolates suspected as AmpC β - lactamase producers.

All cefoxitin resistant *K. pneumoniae* isolates were screened for AmpC β - lactamase production by confirmatory method, the modified three -dimensional test. Results revealed that out of 17 cefoxitin resistant *K. pneumoniae* isolates , AmpC β - lactamase production was confirmed by MTDT in 3 (17.6%) isolates (Table 4-14) .

Table (4-14):Numbers and percentages of AmpC β -lactamase producing carbapenemase positive *Klebsiella pneumoniae* isolates by phenotypic confirmatory methods.

No.(%) of cefoxitin resistant isolates	No.(%) of AmpC β -lactamase producers detected with		No.(%) of non AmpC β - lactamase producer isolates
	MTDT [*]	AmpC disk test	
17	3 (17.6 %)	2 (11.8 %)	12(70.6%)

[*]**MTDT**: Modified three-dimensional test
χ^2 is not significant in ($p < 0.05$)

AmpC β - lactamase production was further confirmed by the AmpC disk test. Indentation of the cefoxitin inhibition zone indicating strong AmpC producer was observed in 2(11.8 %) of isolates, which were also positive by MTDT, whereas flattening (weak AmpC) was not observed in any of cefoxitin resistant isolates (Table 4-14). However ,there is no significant difference ($p<0.05$) between two methods.

4.14. Inducible AmpC β - Lactamase Production

Results of this study revealed that none of cefoxitin resistant isolates gave visible reduction in ceftazidime- imipenem antagonism test, revealing the presence of only plasmid mediated resistance in these isolates.

4.15. Molecular Characterization of AmpC β - Lactamase among Carbapenemase Positive *K. pneumoniae* Isolates.

Detection of AmpC β - lactamase production among carbapenemase positive *K. pneumoniae* isolates was done previously by phenotypic (MTDT and AmpC disk) tests.

All carbaperemase positive isolates were further investigated by PCR for the presence of *bla*AmpC gene. Results demonstrated that only 13 (76.5 %) isolates gave PCR products with AmpC specific primers (Figure 4-18 and Table 4-15). All these isolates showed cefoxitin resistance in cefoxitin susceptibility test (Figure 4-1).

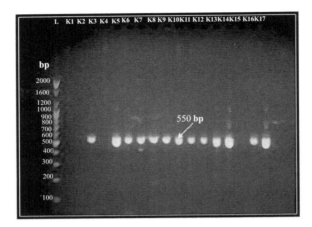

Figure (4-18): Agarose gel electrophoresis (1.5%agarose,70 % volt for 2-3 hrs) for *bla*AmpC **gene product (ampilified size 550 bp) using DNA template of carbapenem-resistant** *K. pneumoniae* **isolates extracted by using salting out method. Lane (L), DNA molecular size marker (100-bp ladder). Lanes (K3, 5, 6, 7, 8, 9, 10, 11, 12, 13, 14, 16 and 17)of** *K. pneumoniae* **isolates show positive results with** *bla*AmpC **gene. Lanes (K1, 2,4 and 15) show negative results with** *bla*AmpC **gene.**

4.16.Combination of Carbapenemases and β - Lactamases Genes

As shown in table (4-15),the distribution of β - lactamases among isolates were so varied. PCR assay revealed that 2(11.8 %),2(11.8%), 5(29.4%), 6(35.5%) and 2(11.8%) isolates carried 1,2, 7, 8,9 carbapenemase , ESBL and AmpC β - lactamase genes, respectively.

Table (4-15): Distribution of *bla* genes among carbapenemase genes harboring *K. pneumoniae* isolates (n=17)

Isolate symbol	Carbapenemase genes						ESBL genes							AmpC gene	Total genes
	blaKPC	blaIMP	blaVIM	blaNDM-1	blaOXA-23	blaSME	blaTEM	blaSHV	blaCTX-M	blaOXA-1	blaPER	blaVEB	blaGES	blaAmpC	
K1	-	-	-	-	+	-	-	-	-	-	-	-	-	-	1
K2	-	-	-	-	+	-	-	-	-	-	-	-	-	-	1
K3	-	-	+	-	+	-	+	+	+	+	-	-	-	+	7
K4	-	-	+	-	+	-	-	-	-	-	-	-	-	-	2
K5	-	-	+	-	+	-	+	+	+	+	-	-	-	+	7
K6	-	-	+	-	-	-	+	+	+	+	+	-	-	+	7
K7	-	-	+	-	+	-	+	+	+	+	+	-	-	+	8
K8	-	-	+	-	+	-	+	+	+	+	+	-	-	+	8
K9	-	-	+	-	-	-	+	+	+	+	+	-	-	+	7
K10	-	-	+	-	+	-	+	+	+	+	+	-	-	+	8
K11	-	-	+	-	+	-	+	+	+	+	+	-	-	+	8
K12	-	-	+	-	+	-	+	+	+	+	-	-	-	+	7
K13	-	-	+	+	+	-	+	+	+	+	+	-	-	+	9
K14	-	-	+	+	+	-	+	+	+	+	+	-	-	+	9
K15	-	-	-	+	+	-	-	-	-	-	-	-	-	-	2
K16	-	-	+	-	+	-	+	+	+	+	+	-	-	+	8
K17	-	-	+	-	+	-	+	+	+	+	+	-	-	+	8
Total	0	0	14	3	15	0	13	13	13	13	10	0	0	13	

4.17. Transfer of Carbapenem Resistance

The natural ability of carbapenemase genes to disseminate into other bacterial strains was examined by conjugation experiments using two carbapenemase positive *K. pneumoniae* (K13 and K14) isolates as donor strains and rifampicin – resistant *E. coli* MM294, as a recipient strain .

Results of conjugation experiments for carbapenemase harboring *K.pneumoniae* isolates were shown in Table (4-16).Conjugation experiments and PCR assay confirmed that the *K. pneumoniae* isolates K13 and K14 were able to transfer *bla*VIM genes in transconjugants, Figure (4-19) .However, *K.pneumoniae* possessed *bla*NDM-1and *bla*OXA-23 can not be transferred by conjugation.

Table (4-16): Characterization of carbapenemase- positive *K. pneumoniae* isolates and their transconjugant.

Isolate	Genotypic Character	MIC (µg /ml)	
		IMP ($\geq$16)	**MEM** ($\geq$16)
K. pneumoniae K13 (clinical isolate)	*bla*VIM, *bla*NDM-1,*bla*OXA-23,*bla*TEM, *bla*SHV,*bla*CTX-M,*bla*OXA-1,*bla*PER,*bla*AmpC	>32	>32
E. coli transconjugant (TCK13)	*bla*VIM, *bla*SHV, *bla*CTX-M, *bla*AmpC	>32	>32
K. pneumoniae K14(clinical isolate)	*bla*VIM, *bla*NDM-1,*bla*OXA-23,*bla*TEM, *bla*SHV,*bla*CTX-M,*bla*OXA-1,*bla*PER,*bla*AmpC	>32	>32
E. coli transconjugant (TCK14)	*bla*VIM, *bla*SHV, *bla*CTX-M, *bla*AmpC	>32	>32
*E.coli*MM294 recipient		0.002	0.008

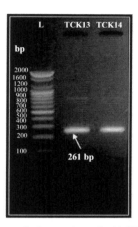

Figure (4-19): Agarose gel electrophoresis (1.5%agarose,70 % volt for 2-3 hrs) for *bla*VIM **PCR product (ampilified size 261 bp) related to the transconjugant** *E.coli* **isolates TCK13 and TCK14. Lane (L) , DNA molecular size marker (100-bp Ladder). Lanes (TCK13 and TCK14) show positive results with** *bla*VIM **gene.**

The conjugation frequencies (the number of transconjugants divided by the number of recipient cells) ranged from 4.6×10^{-2} in *K.pneumoniae* K13 to 6.5×10^{-2} in *K. pneumoniae* K14 transconjuate as shown in Table (4-17).

Table (4-17): Results of conjugation between donor (K13,K14)   and recipient(MM294) *E. coli* **isolates .**

Type of donor isolate	Total No. of isolates		Conjugation frequency
	Transconjugant	**Recipient**	
K. pneumoniae K13	46×10^{2}	1×10^{5}	4.6×10^{-2}
K. pneumoniae K14	130×10^{2}	2×10^{5}	6.5×10^{-2}

Antibiotic susceptibilities for the two transconjugates TCK13 and TCK14 were also detected .Results showed that the MICs for imipenem and meropenem were much higher (>32 μg /ml) relative to those of the recipient (Table 4-16 and Figure 4-20).

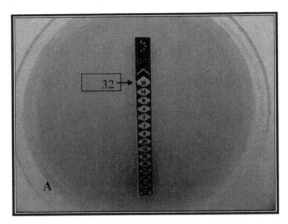

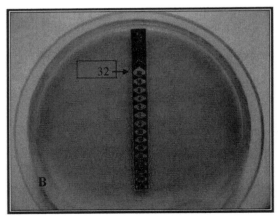

Figure (4-20): MICs by M.I.C.E test for transconjugant *E.coli* isolate. (A),MIC for IMP >32 μg /ml and (B), MIC for MEM >32 μg /ml. IMP, imipenem; MEM, meropenem .

By PCR ,results showed that the *K.pneumoniae* K13 and K14 isolates were able to transfer the *bla*SHV, *bla*CTX-M and *bla*AmpC in transconjugants Table (4-16),Figures(4-21),(4-22),and (4-23).

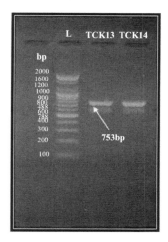

Figure (4-21): Agarose gel electrophoresis (1.5%agarose,70 % volt for 2-3 hrs) for *bla*SHV PCR product (ampilified size 753 bp) related to the transconjugant *E.coli* isolates TCK13 and TCK14. Lane (L) , DNA molecular size marker (100-bp Ladder). Lanes (TCK13 and TCK14) show positive results with *bla*SHV gene.

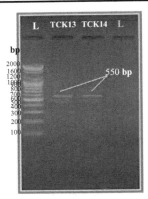

Figure (4-22): Agarose gel electrophoresis (1.5%agarose,70 % volt for 2-3 hrs) for *bla*CTX-M **PCR product (ampilified size 550 bp) related to the transconjugant** *E.coli* **isolates TCK13 and TCK14. Lane (L) , DNA molecular size marker (100-bp Ladder). Lanes (TCK13 and TCK14) show positive results with** *bla*CTX-M **gene.**

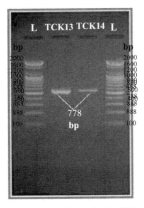

Figure (4-23): Agarose gel electrophoresis (1.5%agarose,70 % volt for 2-3 hrs) for *bla*AmpC **PCR product (ampilified size 550 bp) related to the transconjugant** *E.coli* **isolates TCK13 and TCK14. Lane (L) , DNA molecular size marker (100-bp Ladder). Lanes (TCK13 and TCK14) show positive results with** *bla*AmpC **gene.**

113

5.1.Survey of *K. pneumoniae* Isolates

Results of the present study revealed the presence of 177 (22%) isolates belonged to *Klebsiella* spp. (Table 4-1) .In a previous study in Hilla by Al- Charrakh (2005) , 29(13.8%) *Klebsiella* spp. isolates were obtained from 209 clinical samples. Another study by Al-Muhannak (2010) reported that *Klebsiella* spp. was the second frequently 62 (30.5%) isolated organisms in Najaf hospitals. In Indonesia, Radiji *et al.* (2011) documented 17.7% prevalence rate of *K.pneumoniae.*

Podchun and Ullmann (1998) documented that the principle pathogenic reservoirs for transmission of *Klebsiella* are the gastrointestinal tract and the hands of hospital personnel which increase the likelihood of person – to -person transmission , and contaminated equipments are also important factors promoting the spread of *Klebsiella* spp.

Results from Table (4-1) showed that 117 (14.6%) isolates were identified as *K. pneumoniae* .This result is in agreement with a previous local study in Hilla by Al-Saedi (2000) who found that *K. pneumoniae* isolates comprised (15.3%) from 725 clinical samples . In another study, Al-Sehlawi (2012) reported that the detection rate of *K. pneumoniae* was (14%) among all pathogens isolated from clinical samples in Najaf hospitals. Acheampong *et al.* (2011) recorded 9.3% prevalence rate of *Klebsiella* spp. with *K. pneumoniae* being the highest recovered species (74.4)% followed by *Klebsiella oxytoca* (24.1%) .

However, prevalence and distribution of *K.pneumoniae* varied among Hilla hospitals , Table (4-1) shows distribution of *Klebsiella pneumoniae* in Hilla hospitals: Babylon teaching (23%) , Chest diseases centers (14.8%), Al-Hilla teaching (11.1%) and Merjan teaching (3%).This variation among hospitals related to specification of each hospital for receiving patients suffering from specific diseases related to those hospitals .High prevalence rate (23%) was detected in Babylon teaching hospital, this range may be related to that most patients received were infants and premature babies suffering from diarrhea, vomiting ,meningitis and premature neonates in ICU for long period ,followed by Chest Diseases Center (14.8%), which specified for examining outpatients sputum with respiratory problem most of them for diagnosis of tuberculosis. High prevalence rate may be related to that most patients were debilitated by other diseases like diabetes mellitus and bronchoplumonary diseases .Livneson and Jawetz (2000) mentioned that the oragansim is carried on respiratory tracts of about 10% of normal people ,who are borne to pneumonia if host defense are impaired .

Results of the present study revealed that 569/801(71%) of other bacterial spp. isolates were recovered from clinical samples, (Table 4-1).There is no doubt that hospitals are typical environments for the presence of pathogens such as *S.aureus E. coli*, *Klebsiella*, *Proteus*, *Morganella*, *Enterobacter*, *Citrobacter*, *Serratia*, *Acinetobacter* and *Pseudomonas* spp. (Kucukates and Kocazeybek, 2002 and Azimi *et al.*,2011).However, the dissemination of bacterial isolates in clinical samples may be due to their ability to cause different nosocomial infections and resistance to a wide range of antibiotics.

K. pneumoniae are Gram- negative bacteria which are part of the normal human intestinal flora and are frequently spread via fecal-oral contamination. Pathogenic isolates can be carried in the gut for years in healthy adults and only emerge when intestinal conditions change (Schwaber and Carmeli,2008). The results of Table (4-2) showed that the majority of *K. pneumoniae* isolates 38/141(27%) were obtained from stool samples. High prevalence of *K. pneumoniae* in stool samples was demonstrated by other researchers , Al-Saedi (2000) in Hilla , (14%) , Ali *et al.* (2010) in Jordon, (20%) and Sarojamma and Ramakrishna (2011) in India, (50%).

From the perspective of the world community, acquired lower respiratory tract infections are an important cause of morbidity and mortality for all age groups. Each year, approximately 7 million people die as a direct consequence of acute and chronic respiratory infection (WHO ,1995). Lower respiratory tract infections are very common, with an incidence in the world population of 40-50 per 1000. Since the etiology agents of lower respiratory tract infections cannot be determined clinically, microbiological investigation is critical for both treatment and epidemiological purpose (Ozyilmaz *et al.*, 2005).

In sputum , *K.pneumoniae* was detected in 19/128 (15%) of samples, (Table 4-2). Increasing prevalence of *K.pneumoniae* in sputum was observed by other researchers ,Al- Muhannak (2010) , (15.7%) and Al-Sehlawi (2012), (16%).

The frequent cause of vaginal discharge is an infection or colonization with different microorganisms (Mylonas and Friese, 2007).Vaginitis, whether infection or not, poses one of the most common problems in gynaecology, and is one of the main reasons leading the females to seek advice from a physician approximately 10 millions office visits annually (Kent,1991;Syed and Braverman.,2004). As shown in Table (4-2), 18/116 (15. 5%) isolates were obtained from vaginal swabs. In Pakistan, Mumtaz *et al.* (2008) recorded that 17 (10.5%) of *K.pneumoniae* were recovered from vaginal swabs. However, most microbes have an optimal pH range in which they show an improved activity. Any intervention with the pH of the system may result in the growth of other microbes. The same phenomenon is also applicable for the vaginal system, which is widely populated with the lactobacilli species. The pH is mainly maintained by the production of lactic acid by the lactobacilli (Kumar *et al.*,2011).When the pH is changed due to insertion of contraceptive devices, pregnancy and intercourse ,this may result in the lactobacilli population and a subsequent increase in the growth of other microbes (Aroutcheva *et al.*,2001).

Urinary tract is the common site of infection by *Klebsiella* accounts (6-17%) from all nosocomial urinary tract infections (Podschan and Ullmann,1998) and according to Najmadeen (2006) ,nosocomial urinary tract infections accounted (24%). Results from Table (4-2) revealed that 10/97 (10%) *K.pneumoniae* isolates were obtained from urine samples. In Najaf, Hadi (2008) found that *K.pneumoniae* comprised 23.1% of all nosocomial pathogens obtained from patients with significant bacteruria. However , one of the most common reasons of urinary tract infections is the transmission of bacteria from the gastrointestinal tract to the anterior of urethra , especially in the cases of women .In addition , the secreted vaginal fluids nearby urethra, which create a suitable condition , may help in the presence of bacteria predominantly (Wing *et al.*, 1999) . Moreover , the urine is an appropriate media for the growth and multiplication of bacteria that cause those infections (Bartges, 2007) , especially , in the case of infections associated with the use of the urinary catheter techniques (Godfrey and Evans ,2000).

A wound is a break in the skin and exposure of subcutaneous tissue following loss skin integrity that provides a moist worm, and nutritive environment that is conductive to microbial colonization and proliferation (Bowler *et al.*, 2001) . In the present study *K.pneumoniae* were detected in 8/60(13.3%) of the wound samples. Another study conducted by Achempong *et al* .(2011) recorded that 12.6% of *K.pneumoniae* were recovered from wound infections.

Regarding burn infections, 18/153 (11.7%) of samples were positive for *K.pneumoniae* .In Sulaimani, Najmadeen (2006) found that prevalence rate of *K.pneumoniae* was (30.7%) from burn infections. *K.pneumoniae* is an opportunistic pathogen found along with other bacteria as part of transient normal flora of the human skin, when the host is immunocompromized,as in the case of a thermal burn or a surgical wound (Obiajuru *et al.*,2010).These opportunistic bacteria can quickly colonize and infect the burn or wound sites.

In blood , prevalence rates of *K.pneumoniae* was 3/58(5%), (Table 4-2) . In Spain , a study performed by Rubio- Perez *et al.* (2012) showed that *K.pneumoniae* is the second causative agents (10%) recovered from blood specimens.

From ear samples *K.pneumoniae* comprised 2/30 (6.6%), (Table 4-2).This result is higher than that recorded by Rampure *et al.* (2013) who found that (1.8%) *K.pneumoniae* were detected in ear swabs.

Results of the present study indicated that only 1/8 (12.5%) isolate of *K.pneumoniae* was obtained from eye samples .In Jordan, a study performed by Al-Shara (2011) demonstrated that prevalence rate of *K.pneumoniae* isolates from eye infections was (4.2%) .

Results also showed that no *K.pneumoniae* isolates were recovered from throat samples (Table 4-2), and this result could be due to limited number of throat samples taken in this study. However, In Poland Sekowska *et al.*(2011) found that the detection rate of *K.pneumoniae* from throat swabs was (32%).

In this study, VITEK 2 system was used to confirm identification of *Klebsiella* to species and subspecies levels and to avoid variability in findings of biochemical tests. Results of the present study indicated that *K.pneumoniae* subsp. *pneumoniae* was the most frequent subspecies 77/117(9.6%), followed by *K.pneumoniae* subsp.*ozaenae* 34/117(4%) and *K.pneumoniae* subsp. *rhinoscleromatis* 6/117 (1%) (Table 4-2). Dominance of *K.pneumoniae* subsp. *pneumoniae* among all other subspecies was supported with a report documented by Al-Sehlawi (2012) who stated that *K.pneumoniae* subsp. *pneumoniae* was the most frequent occurring subspecies., accounting for 88.9%.

Results showed that 26/141 (18%) of *K.pneumoniae* subsp. *pneumoniae* ,8/141 (6%) of *K.pneumoniae* subsp.*ozaenae* and 4/141(3%) of *K.pneumoniae* subsp. *rhinoscleromatis* were obtained from stool samples .Similar findings were recorded by Al-Charrakh (2005) who found that most *Klebsiella* spp. were recovered from stool samples and *K.pneumoniae* subsp *pneumoniae* was the most frequently 87% occurring subspecies followed by *K.pneumoniae* subsp.*ozaenea* (9.5%) and *K.pneumoniae* subsp. *rhinoscleromatis* (3.5%).

In sputum samples, the prevalence rate was 8/128 (6%) for *K.pneumoniae* subsp.*pneumoniae* , 9/128 (7%) for *K.pneumoniae* subsp. *ozaenae* and 2/128 (2%) for *K.pneumoniae* subsp. *rhinoscleromatis*. Al-Muhannak (2010) pointed out that the detection rates of *K.pneumoniae* subsp. *pneumoniae*, *K.pneumoniae* subsp. *ozaenae* and *K.pneumoniae* subsp. *rhinoscleromatis* in patients with lower respiratory tract infections in Najaf were (4.3%),(8.6%),and (8.6%), respectively.

5.2.Detection of β -lactam Resistant Isolates

Resistance to β-lactam antibiotics is now a problem in patients throughout the world. The prevalence of β-lactamases among clinical isolates vary greatly worldwide and in geographic areas and are rapidly changing over time. Compared to other β - lactam antibiotics, ampicillin and amoxicillin are the most commonly used antibiotics in the therapy of bacterial infections and can provide a comprehensive primary screening of β - lactam resistant isolates, because the isolates that is resistant to carbenicillin and cephalosporins , is already resistant to ampicillin and amoxicillin (Bush *et al.*, 1995).

As shown in table (4-3) , 91/117 (78%) of *K. pneumoniae* isolates were resistant to ampicillin and amoxicillin. This result is in accordance with a previous study in Hilla by Al- Charrakh (2005) who stated that 73.8% *Klebsiella* isolates obtained from clinical samples were resistant to both ampicillin and amoxicillin. In Najaf, Al- Muhannak (2010) found that 98.2% of *K. pneumoniae* were resistant to both antibiotics .

In this study the high percentage of resistant for these antibiotics could be attributed not only to the production of β - lactamases, but also other resistance mechanisms like decrease the affinity of target PBPs or decrease permeability of the drug into the cell (Jacoby and Munoz-Price, 2005). Amyes (2003) mentioned that there are three further resistance mechanisms include conformational changes in PBPs, permeability changes in the outer membrane, and active efflux of the antibiotic. Other studies reported that *qnr* genes (integron-associated) are associated with resistance to several classes of antibiotics including β-lactam (Paterson, 2006).

5.3.Antibiotic Susceptibility Profile

Antibiotic resistance has become a major clinical and public health problem. The increasing use of antimicrobials in humans, animals and agricultures has resulted in many pathogens developing resistance to these powerful drugs (Sakharkar *et al* .,2009).

Results from Figure (4-1) revealed that higher resistant rate was found for carbenicillin (99%) , ampicillin (94.5%) , piperacillin (82.4 %) . This result in agreement with a pervious study in Hilla by Al- Asady (2009) who found that all 15 (100%) β-lactam resistant *Enterobacteriaceae* isolates were resistant to ampicillin , piperacillin and carbencillin . Al –Hilli (2010) stated that all *K.pneumoniae* isolates were resistant to carbenicillin (100%) and (81%) to piperacillin . High resistance to this class of antibiotics may be due to widespread use of these antibiotics in Hilla hospitals.

The present study showed a high level of resistance to cephalosporins : ceftazidime (86.8%), cefotaxime (83.5%) , ceftriaxone (82.4%) and (80.2%) for cefepime, there is also a wide range of resistance to azteroanam (79%). Cai *et al* .(2011) reported that resistance rate of *K.pneumoniae* isolates to ceftazidime , cefotaxime and cefepime were (70.59%), (88.24%) and (64.71%) respectively. A report by Patel *et al.* (2012) documented that susceptibility to ceftriaxone and cefotaxime was equal , (50%) in *K.pneumoniae* isolates collected from Indian neonates.

High level of resistance to third generation cephalosporins could be attributed to the production of ESBLs,since it mediates resistance to broad spectrum cephalosporins (e.g. ,ceftazidime ,ceftriaxone and cefotaxime) and aztreonam (Umbadevi *et al.,*2011).

In the present study there was a high level of resistance (81.3%) to β-lactam /β -lactamase inhibitor combination (amoxicillin / clavulanic acid). This is likely due to the heavy selection pressure from overuse of this antibiotic and seem to be losing the battle. A similar result was recorded by Al-Shara (2011) who noted that 78.5% of *K.pneumoniae* were resistant to amoxi-clav acid.

In spite of the restricted use of cefoxitin in treatment of bacterial infection in Iraq , results of this study revealed a higher resistance (78%) to cefoxitin among clinical isolates of *K.pneumoniae*. This result is in accordance with the findings recorded by other researchers, Al-Muhannak (2010),(78.7%) and Al-Sehlawi (2012),(70.9%) in Najaf. Cefoxitin antibiotic is stable to broad spectrum β – lactamase mediated hydrolysis as compared with other cephalosporins .The resistance to cefoxitin may be as a result of the development of porin –deficient mutants (Manchanda and Singh, 2003). In addition , increasing numbers of bacterial strains express different types of β- lactamases including inducible and/ or plasmid mediated AmpC type of enzyme may also increase the chance for resistance to cefoxitin (Siu *et al.*, 2003).

There is reduced activity of cefaclor and cefprozil antibiotics with resistance rate of 79.1% each .In a related study in Hilla, Al-Asady (2009) mentioned that all *K.pneumoniae* isolates were resistant to both antibiotics.

Results from Figure (4-1) showed that resistance rate to imipenem was (10 %). In spite of the low level of resistance, this result is higher than that reported by other local studies contacted in Iraq which reported that the susceptibility of *K.pneumoniae* isolates collected from clinical and environmental samples to imipenem was (100%)   (Hadi,2008;Al-Asady ,2009 ; Al- Muhannak , 2010 and Al-Hilli,2010). While in Najaf , Al-Sehlawi (2012) found that only four isolates (3.9%) of *K.pneumoniae* were resistant to imipenem and meropenem antibiotics .Pathak *et al.*(2012) demonstrated 2% resistance to imipenem by *K.pneumoniae* in a surveillance study in two hospitals in India. In another study Shahid *et al.*(2012) reported that the resistance to imipenem was 1.8(16/893) among *E.coli* and *K.pneumoniae* isolates. Reasons behind resistance may be due to inappropriate duration of antibiotic therapy and subtherapeutic concentrations of the drug (Baquero *et al.*,1997 and Philippe *et al.*,1999)

The present study revealed that the resistance against meropenem (17.6%) was more than imipenem . Meropenem is well –tolerated and offers several potential advantages , including greater *in vitro* activity against Gram –negative pathogens and the option of bolus administration (Verwaest *et al.*, 2000) .Beside these , problem of renal metabolism of imipenem , and risk of seizures (Prakash, 2006), and availability of meropenem only in Hilla hospitals might be the reasons behind possible grater use of meropenem over imipenem and hence the high prevalence of resistance.

However , different frequencies of resistance of *K.pneumoniae* to meropenem were reported in different countries (7.5%) in Pakistan (Amin *et al.* , 2009) ; (43.6%) in South India (Parveen *et al.*, 2010b) ; (5.4%) in Indonesia (Radji *et al*,. 2011) and (12.8 %) in Nigeria (Ejikeugwu *et al.*, 2012).

Regarding resistance to ertapenem , the resistance rate was (18.7%).Ertapenem is the least active carbapenem against most strains producing carbapenemase and therefore the first marker that indicates the likelihood of carbapenemase occurrence (Overturf , 2010 and Thomson , 2010). Specificity is questioned because enterabacteria with ESBL and porin mutations are also resistant to ertapenem (Flonta *et al.*, 2011)

A recent study conducted in Nigerian university hospital found 100% susceptibility to ertapenem and imipenem among ESBL – producing *K.pneumoniae* (Igbinoba and Osazuwa, 2012).

Results of the present study revealed that amikacin was more effective (28.6 %) than other aminoglycosides, kanamycin (50%) and gentamicin (40.6%).High efficiency of amikacin may be due to its less vulnerability to bacterial enzymes than other aminoglycosides . In both *Enterobacteriaceae* and other Gram-negative rods, gentamicin and aminoglycosides resistance is often due to the expression of a variety of modifying enzymes including aminoglycoside modifying enzymes(AME),acetylases ,phosphorlyases and adenylases which can impair the effectiveness of antibiotics. Other resistance mechanisms include changes in bacterial membrane permeability and altered ribosomal proteins (Barros *et al.*,1999) .

As shown in Figure (4-1) different levels of resistance were observed to quinolones antibiotics , nalidixic acid (42.8 %) ,ciprofloxacin (33 %) and levofloxacin (28.5%) .Quinolone resistance is typically encoded chromosomally .In this study, resistance against fluroquinolones may reflect antibiotic pressure in Hilla hospitals. In Indonesia , Radji *et al.*, (2011) reported high resistance to levofloxacin (62.2%) and ciprofloxacin (46.9%) among clinical isolates of *K.pneumoniae*.

Quniolone resistance results from mutations in the chromosomally – encoded type II topoisomerases, and via the upregulation of efflux pumps, or point- related genes (Drlica and Zhao,1997 and Tran *et al.*, 2005). The plasmid *qnr* genes play an emerging role in the dissemination of fluoroquinolone resistance (Buchakouer *et al.*,2010).

Percentages of resistance of isolates to the remaining antibiotics were as follows : tetracycline (62.6%), doxycycline and nitrofurantoin (59.3%) each ,trimethoprim-sulfamethoxazole (56%) and chloramphenicol (39.6%).This may be due to multiresistance plasmid harboring *K.pneumoniae* (Tailor,2011).

The high levels of resistance to antibiotics in the present study may be as a result of both intrinsic and acquired mechanisms .The resistance is widespread and constitutes serious clinical threat (Mathur *et al.*,2002).In addition ,the selection pressure of antibiotics in hospital environment lead to multiple resistance to these drugs .Zakaria (2005) mentioned that inappropriate and incorrect administration of antimicrobial agents and lack of appropriate infection control strategies may be the possible reasons behind increasing resistant rate of *K.pneumoniae* to common used antimicrobial drugs.

5.4. β - lactamase production

β - lactamase production in 91 β - lactam resistant *K. pneumoniae* isolates was detected by using rapid iodometric method. Table (4-4) showed that the majority 77/91 (84.6%) of tested isolates were positive with the rapid iodometric test, which could be attributed to the production of plasmid – mediated or chromosomally encoded β - lactamases.

In a related study in Hilla, Al-Hilli (2010) reported that out of 25 β - lactam resistant *K. pneumoniae* isolates, 16 (64%) were able to produce β - lactamases. In Najaf, AL- Sehlawi (2012) pointed out that 58 (61.1%) of clinical isolates of *K. pneumoniae* were detected as β - lactamase producers.

However, Table (4-4) revealed that 14/91(15.4%) isolates were negative with rapid iodometric method, suggesting that these isolates either producing these enzymes with low quantities or they have no β - lactamases. Babini and Livermore (2000) mentioned that most *K.pneumoniae* isolates produce class A chromosomally mediated SHV- 1 β - lactamases , these enzymes are constitutive and are usually produced at low or moderate levels but are sufficient to protect against ampicillin , amoxicillin, carbenicillin and ticarcillin.

5.5.Carbapenemase Production

Carbapenems are a group of β -lactams which consists of imipenem,meropenem,doripenem ,ertapenem,panipenem and biapenem. For more than two decades, carbapenems have been considered the last line of therapy for multidrug-resistant isolates that are prevalent in many Gram-negative bacterial species ,especially those producing ESBLs or/derepressed AmpC β -lactamase (Queenan and Bush,2007 and Peirano *et al* .,2009).

However , more recently carbapenem –resistant *K.pneumoniae* have emerged in United States and various parts of the world (Woodford *et al*.,2004;Peirano *et al* .,2009;Aktas *et al*.,2012 and Sacha *et al*.,2012).This species is resistant to almost all available antimicrobial agents, and infections with this organism have been associated with high rates of morbidity and mortality, particularly among persons with prolonged hospitalization and those who are critically ill and exposed to invasive devices (e.g., ventilators or central venous catheters) (Wachino, *et al*.,2004 and Schwaber and Carmeli,2008).The main mechanism of resistance to carbapenems in *K.pneumoniae* is through the production of a carbapenemase , although resistance is not limited to this mechanism solely . Another method of resistance includes ESBLs and or / AmpC production coupled with outer membrane porin (OMP) alterations (Thomson,2001).

Results revealed that out of 91 –β lactam resistant isolates,17 isolates were found to be resistant to carbapenem antibiotics by standard disk diffusion method, 9/91(10%) were resistant to imipenem ,16 /91(17.6 %) were resistant to meropenem and 17/91 (18.7%) were resistant to ertapenem (Table 4-6 and Figure 4-1). However, many previous studies conducted in Iraq failed to detect resistance to carbapenem in *K.pneumoniae* (Hadi,2008; Al- Asady ,2009; Al-Muhannak , 2010 and Al- Hilli, 2010).

All carbapenem –resistant isolates were screened by three phenotypic tests for carbapenemase production. Table (4-7) shows that 11(65%) of the isolates gave positive results by imipenem- EDTA disk test . Different studies which have used the IMP-EDTA to detect MBL production in *K.pneumoniae* reported a very wide range of prevalence varying from

2.97% to 50% (Deshmukh *et al.*,2011 and Al-Sehlawi,2012). However, there are six isolates which gave negative results with EDTA disk synergy test.This means that EDTA may not inhibit the activity of all β-lactamases suggesting the absence of a class B1 enzyme (Like IMP and VIM) or these isolates may produce other enzymes (Like IMI,GES and KPC) that were not inhibited by EDTA .Moreover , they may have other mechanisms of carbapenem resistance like hyperproduction of AmpC β-lactamases associated with a loss of outer membrane proteins , efflux pumps, and mutations that alter the expression and /or function of porins and PBPs (Cao *et al.*,2000 and Papp-Wallace *et al.*,2011)

The most easily performed test for KPCs is the modified Hodge's test (Figure 4-5), which has been found to be 100% sensitive for the detection of the carbapenemase , although not specific for KPC producers (Anderson *et al.*, 2007 and Chen *et al.*, 2012). From Table (4-7) out of the 17 carbapenem- resistant isolates which were enrolled in this study,14 (82%) isolates were found to produce the carbapenemase enzyme by MHT and all the remaining isolates were found to be carbapenemase negative . In a previous local study, Al- Hilli (2010) reported that 1 (2.4%) of *Klebsiella* spp. isolates recovered from Merjan hospital environment was confirmed as carbapenemase producer using modified Hodge's test, whereas the same isolate gave negative result with imipenem- EDTA synergy test .Another study from Pakistan has shown that 30% of *K.pneumoniae* were positive for carbapenemase production by MHT (Amjad *et al.*, 2011) .However, there have been reports of false positive results ,mostly generated by CTX-M producing strains with reduced outer membrane permeability, and some investigators have raised the problem of difficulties in the interpretation of the clover leaf test for weak carbapenemase producers ,particularly for MBLs *Enterobacteriaceae* (Pasteran *et al.*,2009) .

All 17 carbapenem –resistant isolates were tested further by KPC CHROMagar technique to select the carbapenemase- producers *K.pneumoniae* on medium that enhanced by KPC supplement , this medium allows the growth of *K.pneumoniae* carbapenemase producers only and inhibit the KPC negative bacteria . However , results revealed that all screened 17 (100%) isolates were able to give overnight heavy growth on this medium (Table 4-7 and Figure 4-6) .

In Greece, Panagea *et al.* (2011) reported that CHROMagar KPC is a very useful screening medium both for KPC and VIM carbapenemase – producing *Enterobacteriaceae* in stool samples .CHROMagar KPC efficiently identifies colonized patients in a much shorter time, thus permitting immediate implementation of infection control measures to prevent further dissemination , and in case of infection directs therapy away from β -lactam antibiotics.

5.6.Molecular Detection of Carbapenemases in Clinical Isolates of *K.pneumoniae*

Carbapenemases represent the most versatile family of β - lactamases, with a breadth of spectrum unrivaled by other β-lactam – hydrolyzing enzymes. They confer resistance to virtually all β -lactam agents, including penicillins, cephalosporins , monobactams and carbapenems (Queenan and Bush,2007). Carbapenem resistance in *K.pneumoniae* can be mediated by class A carbapenemases (Peirano *et al.*, 2009) , class B metallo-β - lactamases (Limbago *et al.*, 2011) or rarely by class D OXA- type carbapenemases (Walther- Rasmussen and Hoiby, 2006).

To the best of the researcher's knowledge, until now no published research has described the prevalence of carbapenemases among clinical isolates of *K.pneumoniae* in Hilla hospitals. Hence , this study was conducted to detect the distribution and diversity of carbapenemase genes within these isolates using PCR experiments .Six types of carbapenemase genes namely, *bla*KPC, *bla*IMP , *bla*VIM, *bla*NDM-1, *bla*OXA-23 and *bla*SME were analyzed. Results from Table (4-8) revealed that all carbapenem- resistant *K. pneumoniae* isolates harbored at least one type of three different carbapenemase genes (*bla*VIM, *bla*NDM-1 and *bla*OXA-23).

Among 17 carbapenem –resistant *K. pneumoniae* isolates , 14(82.3%) had a *bla*VIM gene (Table 4-8 and Figure 4-7) . In comparison to another study conducted in India, Shahid *et al.*(2012) reported that among 16 carbapenem- resistant *E.coli* and *K.pneumoniae* , *bla*VIM was detected in only 3(18.6%) isolates, all were *K.pneumoniae* .In another study in Greece, Giakkoupi *et al.* (2009) stated that all *K.pneumoniae* isolates carried *bla*VIM that also produce KPC carbapenemase. Birgy *et al.*(2011) reported the first case of imported health care -associated fecal carriage of an Egyptian infant VIM-1-producing *K.pneumoniae* strain in France. The *bla*VIM – genes are located in class 1 integrons as a gene cassttes and have been identified on plasmid with different replicon types (Carattoli ,2009) , increasing the possibility of dissemination and linkage to other antibiotic resistance genes. However, this constitutes the first report on prevalence and detection of *bla*VIM in Hilla hospitals.

Although VIM genes are most common ,more recent attention has been given to NDM –types MBLs. The NDM-1 is a new molecular class B enzymes that were recently recognized from *K.pneumoniae* isolate from a patient in Sweden who seems to have imported from India. Particularly NDM-1 is endemic to India but due to international travel ,its emerging as an important clinical threat worldwide (Patel *et al.*, 2011).The bacteria with NDM-1 gene are known as superbugs and public health must pay more attention to them (Bonomo,2011).Most infections with NDM-1 producers have been reported in adults, but bla_{NDM-1} in two isolates of *K.pneumoniae* was reported in a neonatal intensive care unit in India as well (Roy *et al.*,2011).In 2010, bla_{NDM-1} and bla_{NDM-2} were found in *A.baumanni* (Pefeifer *et al.*,2011), and in 2011, bla_{NDM-5} was found in *E.coli* (Hornsey *et al.*,2011).More than six different NDM allotypes are known (Fallah *et al.*,2011).

In this study,the overall prevalence of bla_{NDM-1} possessing *K.pneumoniae* was found to be 3 (17.6%) (Table 4-8 and Figure 4-8).In one study, Bora and Ahmed (2012) observed that all *K.pneumoniae* with reduced susceptibility to carbapenem carried bla_{NDM-1} gene. In another study ,among 22 NDM-1 producing *Enterobacteriaceae* ,10 *Klebsiella* spp.were found to be positive for NDM-1 at a tertiary care unit in Mumbi (Deshpande *et al.*,2010). This is the first report on prevalence of NDM-1 in Hilla hospitals.

NDM-1 is carried on plasmid or on chromosomes (Chen *et al.*,2012). The rapid emergence of NDM-1 has been related to a moveable plasmid which can be transferred from one bacteria to another,from man to man and even from country to country (Fallah *et al* .,2011).In recent years, many Iraqi patients were travelled to India and to other countries for medical care purpose which may helped in acquiring this gene.

Oxacillinases are only weakly active against carbapenems and are largely confined to *Pseudomonas* and *Acinetobacter* species and only rarely in *Enterobacteriaceae* (Walther- Rasmussen and Hoiby, 2006). OXA-23 represented a new subset of the OXA family, with the highest amino acid identity to OXA-5 and OXA- 10 at 36% (Queenan and Bush, 2007). It has been identified in outbreaks of carbapenem –resistant *Acinetobacter* in Brazil , Korea and United Kingdom (Dalla- Costa *et al.*, 2003; Jeon *et al* ., 2005 and Turton *et al.*, 2005).

Results of the present study revealed that 15 (88.2%) of carbapenem–resistant isolates harbored *bla*OXA-23 (Table4-8 and Figure4-9). This represents the first report of OXA-23 among clinical isolates of *K.pneumoniae* in Hilla. Recently , *bla*OXA-48 mediated resistance to carbepenem has been reported in *K.pneumoniae* in Turkey (Aktas *et al.,* 2012) .The present study found that the prevalence rate of *bla*VIM and *bla*OXA-23 in carbapenem –resistant *K.pneumoniae* isolates from Hilla hospitals is undoubtedly high.

However, *bla*KPC , *bla*IMP and *bla*SME were not detected among carbapenem –resistant *K.pneumoniae* isolates, which could be due to either the absence of *bla*KPC, *bla*IMP and *bla*SME genes or the presence of other type of gene variants that could not be targeted by the primers used in this study.

5.7.Emergence of Carbapenemases Harboring *K.pneumoniae* Isolates

The epidemiology and clinical data associated with infections caused by carbapenemase harboring *K.pneumoniae* have not been well described.In the present investigation an attempted was made to evaluate the clinical characteristics of the 17 patients in whom carbapenemase positive *K.pneumoniae* were isolated. High rate of carbapenemases harboring *K.pneumoniae* isolates were found among young and middle age groups , 1-20 years (2.8 %) and 21- 40 years (1.9%) (Table 4-9).A recent study from Canada, documented that carbapenem intermediate / resistant (CIR) *K.pneumoniae* isolates were mainly found in the 17 years age group (Tailor,2011).

Table (4-9) showed that females were more exposed to infected by carbapenemase positive *K.pneumoniae* isolates (2.5 %) than males

(1.5%).In United States ,Barykov *et al.*, (2013) reported highest occurrence of carbapenemase positive *K.pneumoniae* among females (66.7%) than males (33.3%).This findings may refer to the difference in personal healthier and educational factors.

In term of hospitalization , results of the present investigation showed that carbapenemase positive isolates occurred predominantly among patients on prolonged hospitalization (2.6%) as compared with out patients (0.6 %).This result is in accordance with that obtained by Perez *et al.* (2010) who noticed that all *K.pneumoniae* carrying *bla*KPC-2 or *bla* KPC-3 were isolated from hospitalized patients .In a recent study ,Teo *et al.*(2012) identified that previous hospital stay is a significant predictor for infection with carbapenem –resistant *Enterobacteriaceae*. Accordingly, a prolonged hospital stay may be prone to nosocomial infections with multidrug – resistant pathogens.

In the present study all clinical samples were obtained from patients had previously received some antibiotics within distinct periods. However , all carbapenemase positive isolates were identified from antibiotic received patients ,Table (4-9). Some researchers ,mentioned that previous exposure to several classes of antimicrobials has been associated with carbapenem resistant *K.pneumoniae* carriage or infection, including carbapenems, cephalosporins , fluoroquinolones and vancmycin (Patel *et al.*, 2008 ; Schwaber *et al.*, 2008; Hussein *et al.*, 2009 and Wiener –Well *et al.*, 2010).In a recent study Patel *et al.*(2011) found that the cumulative number of antibiotic exposure appears to the most important driver of carbapenem –intermediate or –resistant *Enterobacteriaceae* infections.

With regard to the clinical origin , Table (4-10) showed that 9/153 (5.9%) of carbapenemase positive *K.pneumoniae* isolates (6 isolates of

K.pneumoniae subspp.*pneumoniae* and 3 isolates of *K.pneumoniae* subspp.*ozaenae,* Appendix 3) were obtained from burns infections.

High detection rate in burned patients may be related to that most samples were taken from victims with 3rd degree of burn, these victims loses first physical barriers and particularly more exposed and susceptible to life – threatening infections . Other factors such as contact either via the hands of the staff caring for the patient or from contact with unsuitable decontaminated equipments ,the same bath also was used for washing different patients in the same unit, despite of that the main problem related to overuse of different antibiotics for long period of time in order to control infection with multiple drug resistant pathogens. This explanation was supported by Askarian *et al* .(2004) who stated that burn patients are at risk for acquiring infection due to their destroyed skin barrier that provide an excellent culture medium for bacteria. In addition, thermal injury itself leads to suppression of white blood cell function , and suppressed immune system, compounded by prolonged hospitalization and overuse of antibiotics for infection control. Kollef and Fraser (2001) also mentioned that the prolonged hospitalization often harbor antibiotic resistant pathogens.

From wound samples 3/60 (5%) of *K.pneumoniae* isolates (2 isolates of *K.pneumoniae* subspp.*pneumoniae* and 1 isolates of *K.pneumoniae* subspp.*ozaenae,* Appendix 3) were obtained (Table 4-10).The laying of patients in the hospital for long periods and inadequate cleaning of operating wound during hospitalization may be a good reasons for infection with carbapenemase positive isolates .In a recent study Tailor (2011) mentioned that wound infections seems to be the main source for carbapenem reduced susceptible (CRS) and carbapenem intermediate / resistant *E.coli* and *K.pneumoniae* .

5.8.Antibiotic Susceptibility of Carbapenemase Positive *K.pneumoniae* Isolates

Therapeutic options for infections caused by *K.pneumoniae* possessing carbapenemases are limited because these organisms are usually resistant to all beta lactam antibiotics (Queenan and Bush, 2007).This emphasizes the need for detecting carbapenemase harboring isolates in Iraqi hospitals,so to avoid therapeutic failure and nosocomial outbreaks.

In the present study, a high rate of resistance among carbapenemase positive *K.pneumoniae* isolates recovered from clinical samples in Hilla city was documented. Results from Figure (4-10) revealed that all isolates were resistant at a higher rate to ampicillin , piperacillin, carbenicillin , amoxi – clav , cefoxitin, ceftazidime, cefotaxime , ceftriaxone, cefepime , cefaclor, cefprozil, aztreonam, ertapenem, tetracycline, doxycycline , trimethoprim-sulfamethoxazole and chloramphenicol. Based on this observation ,the ability of these isolates for fully resistance to antibiotics could be caused either by predominate exposure of present isolates to suboptimal levels of antibiotics ,prolonged used of broad spectrum antibiotics, exposure to isolates carrying resistance genes ,lack of hygiene in clinical environments and usage of antibiotics in food and agriculture .

Acquiring resistance to a carbapenem might include acquiring resistance to multiple other antimicrobial agents (Walther-Rasmussen and Hoiby,2006). Our data have shown that these isolates were resistant to meropenem and kanamycin (94%each), gantamicin and nitrofurantion (88%, each),ciprofloxacin (82%), amikacin, nalidixic acid and levofloxacin (70%) each, and imipenem (53%), denoting that the plasmid harboring the carbapenem resistant genes harbors additional resistant determinants that may act synergistically and lead to the hydrolysis of carbapenems (Karisik *et al.*,2007).

5.9.Drug Resistance Pattern among Carbapenemase Genes Positive *K. pneumoniae* Isolates

Table (4-11) showed that 11/17(65%) of carbapenemase positive *K.pneumoniae* isolates were resistant to at least 9 different antibiotics classes. An XDR phenotype in *Enterobacteriaceae* is undoubtedly represented by carbapenem resistance which is mainly mediated by MBLs of VIM and IMP types. The vast majority of MBL genes are carried on plasmids as gene cassettes inserted into class 1 integrons and are usually associated with aminoglycosides resistance genes (Walsh, 2008) .This results is higher than that reported by Al-Sehlawi (2012) who stated that 44% of *bla*AmpC positive *K.pnuemoniae* isolates were found to harbor an extensive drug resistant (XDR) .In China , Li *et al.* (2012) pointed that among 223 *K.pneumoniae* isolates only 49 (22%) were found to be XDR isolates.

In this investigation pandrug –resistant *K.pneumoniae* isolates were detected in 6(35%) (Table 4-11).All these isolates were resistant to all antibiotics classes tested .This result was higher than that recorded by Aminizadeh and Kashi (2011)who stated that pandrug –resistant pattern was detected in only 1 (10%) isolate of *K.pneumoniae* in Iran.

Infections by PDR *Enterobacteriaceae* ,although still rare ,have been associated with a high mortality(Souli *et al.*,2008) . In one study,among 28 patients suffering from PDR infections in Greece from January 2006 to May 2007,the attributable mortality 33.3% (Falagas *et al.*, 2008).

However, colistin was used for about two decades after its discovery in 1950, but the reported nephrotoxicity and neurotoxicity led to gradual decrease of its use (Michalopoulos *et al.*, 2005) . The reuse of colistin is associated with a possible significant therapeutic problem , namely the advent of bacteria resistant to all classes of available antimicrobial agents, including the polymyxins .These bacteria are pandrug –resistant . It should be noted that the definition of pandrug –resistant Gram-negative bacteria does not include always testing for colistin in many countries. Hsueh (2002) and Kuo (2003) both reported high mortality rate 60% due to *A. baumannii* infections in Taiwan ,however no colistin was used in the *in vitro* susceptibility testing and ,most importantly the drug was not given to patients. Falagas *et al.* (2005) showed that the isolation of a pandrug –resistant Gram-negative rod from clinical specimens does not necessarily mean a bad outcome .

The present study suggested that the new clonal spread of XDR and PDR of carbapenemase positive *K.pneumoniae* isolates had been documented as first prevalence in Hilla city.

The rates of XDR and PDR isolates observed in this study are alarmingly high. This could cause difficulty in treating *K.pneumoniae* – associated infections since fewer and fewer effective drugs are available for treating those highly drug- resistant isolates.

5.10.Minimum Inhibitory Concentrations (MICs) of Carbapenemase Possessed *K.pneumoniae*

Minimum inhibitory concentration is the lowest concentration of an antimicrobial that will inhibit the visible growth of a microorganism after overnight incubation (Andrews ,2001). In the present investigation , the MIC values of 17 carbapenemase positive *K.pneumoniae* isolates against seven– β -lactam antibiotics (ampicillin, cefotaxime, ceftriaxone, ceftazidime, imipenem and meropenem) were evaluated by HiComb and

M.I.C.E tests. Results from Table (4-12) showed that all carbapenemase positive isolates exhibited high levels of resistance to imipenem , meropenem, ampicillin, cefotaxime, ceftazidime and ceftriaxone with concentrations reached beyond the breakpoint values. Queenan and Bush (2007) mentioned that the first cause for suspicion that a carbapenemase is involved in a clinical infection is an elevated carbapenem MIC. High MIC values for carbapenems may be associated with MBL production, and this may be a useful tool to differentiate from other resistance mechanisms, such as efflux pumps or chromosomal inducible AmpC β-lactamase (Livermore,1995).

However, MICs values for imipenem and meropenem were > 32 μg/ml ,each (Table 4-12). In one study , a collection of five related *K.pneumoniae* strains with VIM-1 gene demonstrated imipenem MICs ranging from 2-64 μg /ml (Loli *et al.*, 2006) . Recently, El-Herte *et al.* (2012) pointed out that the MICs values of imipenem and meropenem for *K.pneumoniae* isolates obtained from two Iraqi patients referred to Lebanon were > 32 μg /ml and > 16 μg /ml ,respectively, both isolates harbored *bla*NDM-1 gene.

For ampicillin the MIC value was >256 μg/ml (Table 4-12).Roh *et al.*(2011) found that the MIC value of ampicillin for KPC-2 producing *K.pneumoniae* was >256 μg/ml by E-test. It was also shown by other studies that TEM or SHV enzymes were detected in 80% of ampicillin resistant Gram-negative bacteria with MICs exceeding 256 μg/ml compared to 1-4 μg/ml in non-β-lactamases producer isolates (Livermore ,1995 and Amador *et al.*,2009)

For cefotaxime, ceftazidime and ceftriaxone the MIC values were >240 μg /ml,> 240 μg /ml ,>256 μg /ml, respectively, whereas K1 and K5 isolates showed MICs of 120 μg /ml, 60 μg /ml, 128 μg /ml ,each (Table 4-12). This result is in agreement with a study conducted by Chi *et al* .(2012) who found that carbapenem –resistant *K.pneumoniae* with Ompk36 loss showed MIC values of > 256 μg /ml for cefotaxime , > 256 μg /ml for ceftiraxone, and > 256 μg /ml for ceftazidime. In another study ,Cao *et al*.(2000) found that the MIC values for cefotaxime and ceftazidime in imipenem -resistant *K.pneumoniae* were> 256 μg /ml each. Bogarets *et al* .(2010) noted that all KPC producing *K.pneumoniae* isolates showed MIC values ≥ 64 μg /ml for cefotaxime.

5.11.Production of Extended Spectrum β -Lactamases

The emerging problem of extended spectrum -beta lactamase producing bacteria has become of great importance during the last decades (Rubio-Perez *et al*.,2012) .Since the first report of an ESBL-producing organism in 1980s,there has been a growing interest due to their wide spread and constant evolution, becoming increasingly resistant to most of the commonly used antibiotics (Gomez *et al*.,2008).However, ESBL detection is not carried out in many microbiology units in developing countries ,including Iraq and this could be due to lack of awareness and lack the resources to conduct ESBL identification.

In this investigation, ceftazidime resistance was used to primarily screening of potential ESBL . However, ceftazidime was chosen to detect ESBL producers for the reason that it is best third- generation cephalosporin substrates for most TEM, SHV and CTX-M derived ESBLs (Livermore and Brown, 2001).Results of the present study revealed that the higher resistance percentage was recorded for ceftazidime (86.8%) than other generation cephalosporins like cefotaxime (83.5%) , ceftriaxone (82.4%) and aztreonam (79%), Figure (4-1). However, ceftazidime is able to expose wide range resistance isolates.

Hence, this antibiotic should be used in early ESBL detection in Hilla laboratories .

Earlier studies on the dissemination of ESBL producing *K. pneumoniae* isolates in Najaf hospitals (using ceftazidime disk) showed that the distribution of ESBL producing isolates during 2008-2010 was 64.3% and 72.1% , respectively (Hadi, 2008 and Al-Muhannak , 2010). In India, Chiangjong (2006) found that 95% of *K .pneumoniae* isolates were suspected ESBL producers by using ceftazidime disk as initial screen test. Recently, Al-Sehlawi (2012) reported   that the use of ceftazidime is the best cephalosporin for primarily detection of ESBLs.

The results revealed that all β - lactam resistant *K. pneumoniae* isolates 91 (100%) were found to be ESBL- producers , initially.

Table (4-5) showed that 81/91(89%) of β -lactam resistant *K.pnenmoniae* isolates were identified as ESBL producers. This may be due to chromogenic character of ESBL CHROMagar and its sensitivity and selectivity which enabled the recovery and presumptive identification of most ESBL- producing *Enterobacteriaceae* within 24hrs, while inhibiting the growth of other bacteria including those carrying AmpC β - lactamase type . This is an important feature of CHROMagar medium , since intrinsic AmpC β -lactamase has no clinical relevance, but often leads to ESBL false positive reading in the classical testing methods (Glupczynski *et al*., 2007 and CHROMagar microbiology / www. CHROMagar.com). Thus ,this medium offers high sensitivity and high specificity combined with a short time to reporting the results , overall, it cost per single test. In Najaf , Al-Sehlawi (2012) reported that 37(47.4%) of the β -lactamase producing *K. pnenmoniae* isolates were confirmed as ESBL producers by this technique.

By return to rapid iodometric test,77/91(84.6%) isolates were detected as β-lactamase producer while the rate increased to 81/91(89%) in case of detecting ESBLs using ESBL CHROMagar technique. Iodometric test represent a primitive detection test as compared with the main advance specific ESBL CHROMagar technique that may explain the high rate of ESBLs positive isolates.

Disk approximation test remains a reliable, convenient , and inexpensive method of screening for ESBLs. However, the interpretation of the test is quite subjective. Sensitivity may be reduced when ESBL activity is very low leading to wide inhibition zones around the cephalosporin and aztreanam (Vercauteren *et al.*, 1997). Result from Table (4-5) revealed that out of 91 β -lactam resistant isolates , only 45(49.5%) were confirmed as ESBL producers by disk approximation test.This result is much higher than that reported by other researchers in Iraq, Al-Charrakh (2005) in Hilla city (10.5%) ; Al-Muhannak (2010) , (21.4%) and Al-Sehalwi (2012),(0%) in Najaf city, but this result correlates with the results being reported by Basavraj *et al.* (2011) who demonstrated the presence of 24 (42.9%) of *K.pneumoniae* isolates as ESBL producers by this test.

However ,ESBL producing *K.pneumoniae* isolates were more frequently detected by CHROMagar than disk approximation method. Disk approximation test lacks sensitivity due to the problem of optimal disk space, the correct storage of the clavulanate containing discs, the in ability of clavulanic acid to inhibit all ESBLs and the inability to detect ESBLs producing isolates which have ability to produce chromosomal and plasmid mediated cephalosporinases (Hemalatha *et al.*, 2007; Fam and El-Damarawy, 2008 and Basavaraj *et al.*, 2011).

The high prevalence of ESBL producing *K.pneumoniae* isolates in this study may be due to the widespread use of the third generation cepholosporins and aztreonam which believed to be the major cause of mutations in the class A TEM and SHV enzymes . Cross- resistance to other unrelated antibiotics may occur and this resistance is transferable to other bacterial species in association with plasmids (Shanmuganathan *et al.*, 2004   and Basavaraji *et al.*, 2011).This study suggests that ESBL producing   *K.pneumoniae* isolates are already endemic in Hilla city. Thus, detection of ESBL production is of importance in hospital isolates . Firstly , these isolates are probably more prevalent than currently recognized .Secondly , ESBLs constitute a serious threat to currently available antibiotics . Thirdly , in situational outbreaks are increasing because of selective pressure due to heavy use of extended – spectrum cephalosporins and lapses in effective control measures.

However ,in this study not all screen positive *K.pneumoniae* isolates were ESBL producers .Thus there may be other mechanisms of resistance to third generation cephaosporins and aztronam .In organisms that produce both ESBL and AmpC ,clavulanate may induce hyperproduction of the AmpC β-lactamase leading to the hydrolysis of the third generation cephalosporins thus masking any synergy arising from the inhibition of ESBL,producing false negative result in the ESBL detection test (Thomson,2001).

5.12.Detection of ESBL Genes among Carbapenemase Harboring *K.pneumoniae* Isolates

Extended spectrum β-lactamase producing *K.pneumoniae* have spread rapidly worldwide and pose a serious threat in heleathcare – associated infections (Kiratisin *et al.*,2008). However,the detection of *bla*TEM *bla*SHV, *bla*CTX-M, *bla*OXA-1, *bla*PER, *bla*VEB and *bla*GES genes were performed with PCR assay .In fact the co-existence of broad spectrum β-lactamase with ESBLs,ESBLs with AmpC,multiple extended with metallo β-lactamase has become common in multiresistance *K.pneumoniae* (Ktari *et al*,2006 and Tailor,2011).

Among different ESBL – genes *bla*TEM were detected in 13/17 (76.5%) of carbapenemase positive *K. pneumoniae* isolates (Table 4-13 and Figure 4-13). The present study demonstrated a notable increase in the frequency of *bla*TEM in Hilla compared with previous study conducted by Al-Asady (2009) who found that 57.1% of *E.coli* and *K. pneumoniae* isolates were positive for *bla*TEM genes . In Iran , Khorshidi *et al*. (2011) stated that 19% of *K. pneumoniae* isolates had *bla*TEM gene. Other study characterized that the frequency of *bla*TEM were (55%) of the confirmed ESBL –producing *K. pneumoniae* (Moosavian and Deiham , 2012).

Reports originated from geographically diverse locations, suggest that local antibiotic usage and practices may play an active role in promoting the selection of point mutations in *bla* TEM-type genes (Vourli *et al*., 2004 and Wachino *et al*., 2004).

Results presented in table (4-13) revealed that 13/17 (76.5%) of carbapenemase positive isolates harbor *bla*SHV gene. The prevalence of *bla*SHV in present study was high compared with the results obtained by Hadi (2008) who revealed that 29.4% of confirmed ESBL producing *K. pneumoniae* isolates had *bla*SHV genes . In India , Jemima and Verghese (2008) reported the prevalence of *bla*SHV as 45% in *Klebsiella* spp., 14% in *E.coli* and 15% in *Enterobacter* spp. Bali *et al* .(2010) identified the presence of SHV-enzyme in 8 (32%) of *K. pneumoniae* isolates in Turkey.

A reports by Tzouvelekis and Bonomo (1999) considered the SHV-1 β-lactamase is most commonly found in *K. pneumoniae* and is responsible for up to 20% of the plasmid-mediated ampicillin resistance in this species .

Plasmid-mediated *bla*SHV genes are possibly mobilized from genome to plasmid mediated by insertion sequence (IS)*26* (Ford and Avison 2004). IS*26* is widely distributed among plasmids facilitating the mobilization of chromosomal sequences containing resistance genes. Also, IS*26* was found associated with a class 1 integrons which is considered as a critical step in the evolution of diverse multiresistance plasmids found in clinical enterobacteria (Miriagou, *et al.* 2005).

In the present study , *bla*CTX-M was detected in 13/17 (76.5%) of carbapenemase positive isolates (Table4-13 and Figure 4-15) .CTX-M first detected in Japan in 1986 from cefotaxime – resistant *E.coli,* which was named FEC-L (Matsumoto *et al.*, 1988). A few years later in 1989 , a similar cefotaxime-resistant *E.coli* strain from Germany was reported to produce β-lactamase enzyme designated CTX-M-1 (Bauernfiend *et al.*, 1990). CTX-M β-lactamases constitute a novel and rapidly growing family of plasmid –mediated ESBLs (Baraniak *et al.*, 2002). In a previous study in Hilla , CTX- M enzymes were identified in 2 (11.1%) of *Klebsiella* spp. recovered from environment of Merjan teaching hospital (Al- Hilli, 2010).

In a survey examined 113 *K. pneumoniae* isolates from five hospitals in three Iranian cities , 16 (23.9%) isolates with CTX-M enzymes have been detected (Ghafourian *et al.*, 2011). In Spain , PCR amplification revealed that out of 162 ESBL- producing *K. pneumoniae* analyzied, 76 (67%) were positive for *bla*CTX-M (Alegria *et al.*, 2011).

In this study there is high prevalence of *bla*CTX-M in Hilla, this may attributed to the pressure from the surrounding by antimicrobial agents such as contact with healthcare facilities .This frequency may be due to insertion sequence such as IS*Ecp1* which appear to enable the mobilization of CTX-M gene (Forssten,2009).

Results also revealed that, 13/17 (76.5%) of carbapenemase positives *K. pneumoniae* carried *bla*OXA-1 gene (Table4-13 and Figure 4-16) .OXA enzymes are regarded as OXA-type ESBls, attack the oxyimino – cephalosporins and have a high hydrolytic activity against oxacillin, methicillin and cloxacillin more than benzylpenicillin (Queenan and Bush, 2007) .The most of OXA derivative genes are plasmid and integron located (Poirel *et al.*, 2001) . The presence of co- resistance gene cassettes on integrons make these genetic elements useful to bacteria by facilitating widespread dissemination through patients from a wide variety of clinical disciplines (Poriel *et al.*, 2002). Low prevalence of *bla* OXA-1 genes has been reported in Najaf, Al- Muhannak (2010) ,6 (9.7%) and Al- Sehlawi (2012) , 2(10%) . While in Hilla city , Al-Hilli (2010) found that all *E.coli* and *Klebsiella* spp. isolates from Merjan teaching hospital were negative in OXA-PCR.

Among the 17 carbapenemase positive *K.pneumoniae* 10 (58.8%) isolates harbored a *bla*PER gene as determined by PCR (Table 4-13 and Figure 4-17) , However this is the first report done with the detection of PER β - lactamase in Hilla . Other studies failed to detect *bla*PER among clinical isolates of *K.pneumoniae* , in Turkey (Erac and Gulay , 2007) , in China (Wei-feng *et al.*, 2009) and in Colmbia (Martinez *et al.*, 2012) , whereas in Korea, one study reported the prevalence of *bla*PER in blood samples from patient with pneumonia (Bae *et al.*, 2011).

In this study , all carbapenemases positive *K. pneumoniae* isolates were negative in VEB and GES PCRs, which could be either due to the absence of *bla*VEB and *bla*GES or the presence of other type of genes that could not be targeted by the primers used in this study. The prevalence of these enzymes is rare compared with TEM, SHV and CTX-M. They are mostly isolated from *P.aeruginosa* and are found in limited geographic region (Thirapanmethee , 2012).

5.13.AmpC β -Lactamase Production among Carbapenemase Positive *K. pneumoniae* Isolates

AmpC β - lactamases, class C classified according to Ambler and to group 1 by Bush- Jacoby -Medeiros , are clinically important cephalosporinases produced by several *Enterobacteriaceae* strains and mediate resistance to cephalothin ,cefalozin,cefoxitin,most pencillins and β -lactam / β -lactam inhibitor combinations (Philippon *et al.*,2002 and Abic *et al.*, 2006) .Although less prevalent than ESBLs (Jacoby, 2009), *bla*AmpC have a comparatively broader substrate range , are not inhibited by traditional β - lactamase inhibitors and bacteria producing them are more likely to become resistant to carbapenems (Philippon *et al.*, 2002).

In this study ,cefoxitin susceptibility of the 17 carbapenemase positive *K.pneumoniae* isolates was tested by standard disk diffusion method (Bauer *et al.*1966). Results of the present study indicated that all 17 (100%) carbapenemase positive *K.pneumoniae* isolates were resistant to cefoxitin (Table4-14) .The frequency of cefoxitin resistance in the present study was higher than that previously recorded by Al-Hilli (2010) who stated that 7(28%) of *K.pneumoniae* were resistant to this antibiotic. Other study characterized that 84 (72%) of *Klebsiella* spp. were found to be cefoxitin resistant in Chennai , India (Subha *et al.*, 2003). Resistance to cefoxitin may be due to over expression of chromosomal *ampC* gene , acquisition of a plasmidic *ampC* gene , porin or permeability mutations , or a combination of these factors (Mulvey *et al.*,2005).

The fact that a standardized phenotypic method for screening and detection of these types of resistance does not exist makes the surveillance and characterization of strains problematic. Screening with cefoxitin disk is recommended for initial detection . However, it does not reliably indicate AmpC production. Some of the phenotypic tests include the modified three dimensional test and AmpC disk test could be used as confirmatory tests.

Thomson and Sanders (1992) first reported three- dimensional test , since then many modification have been made. Coudron *et al.*, (2000) modified original method and demonstrated that the three- dimensional test did not reveal false negative results and only 1 (3.6%) of the 28 AmpC harboring *E.coli* and *Klebsiella* spp . isolates was false positive. Subsequently , Manchanda and Singh (2003) modified the procedure to overcomes all the problems of three- dimensional test.

From Table (4-14) out of 17 cefoxitin resistant *K.pneumoniae* isolates , AmpC β - lactamase production was confirmed in 3 (17.6 %) isolates by using MTDT. In a related study in Hilla by Al- Hilli (2010) , AmpC β - lactamase was produced by MTDT and AmpC disk test in 1 (4%) of 7 cefoxitin resistant *K.pneumoniae* isolates . Al –Sehalwi (2012) estimated that 31 (42.5 %) of *K.pneumoniae* isolates were AmpC β - lactamase producer by using MTDT in Najaf city. In Indian study, plasmid – mediated AmpC β - lactamase production was detected in 149 (79.6%) of screen positive *E.coli* and *K.pneumoniae* by this method. Cefoxitin resistance in AmpC non- producers may be due to some other resistance mechanisms such as , lack of permeation of porins (Pangon *et al.*, 1989) . Another study has demonstrated that the interruption of a porin gene by insertion sequences is a common type of mutation that causes loss or decrease of outer membrane porin expression and increase cefoxitin resistance in *Klebsiella* spp .(Hernandez – Alles *et al.*, 2000).

Simple modifications of MTDT has been introduced , since MTDT is laborious , technically demanding, requiring careful cutting of slit and well , time consuming and needs experience (Rudresh and Nagarathnamma, 2011).

The AmpC disk test is a simple , reliable and rapid method of detection of isolates that produce AmpC β - lactamases, and can be used for routine screening of AmpC enzyme in the routine clinical laboratory (Samatha and Parveen, 2011) .The test accurately distinguished between cefoxitin insusceptibility caused by AmpC production and non β - lactamses mechanisms , such as reduced outer membrane permeability (porin mutations).

Table (4-14) also revealed that out of 17 cefoxitin resistant isolates , AmpC β -lactamase production was detected by AmpC disk test in 2 (11.8%) isolates , which was also showed positive results by MTDT. In a study from India,Samatha and Praveen (2012) characterized that 7 of cefoxitin resistant *K.pneumoniae* gave strong positive result with AmpC disk test. Another study by El- Dien and Kashif (2012) reported that among 139 cefoxitin resistant Gram – negative bacilli , only two isolates of *Klebsiella* spp. were confirmed as AmpC β -lactamase producers by this test .

In this study the prevalence of AmpC β - lactamase was low (if compared with cefoxitin susceptibility results), this may be due to that some isolates having *ampC* genes , but might not be expressed in all the isolates. This means that they might have ' silent genes' or there might be low-level expression of *ampC* genes that was not detected (Jacoby, 2009).

5.14. Inducible AmpC β - Lactamase Production

Many Gram – negative bacteria harbor chromosomal AmpC beta – lactamase genes ,which are constitutively expressed at low level .In general, the expression of chromosomally located *ampC* genes is inducible by beta – lactam antibiotics , such as cefoxitin , cefotetan , and imipenem, and mediated by the regulator AmpR (Polsfuss *et al.*, 2011).

Mutation in the repressor gene *ampD* are the most common cause of constitutive (hyper-) production of AmpC β - lactamases (Schmidtke and Hanson , 2006).As some bacterial species may produce inducible AmpC β -lactamases that can be easily overlooked by routine susceptibility tests, Cantarelli *et al.*(2007), reported a new test to detect and confirm the presence of inducible AmpCs among enterobacterial strains, the ceftazidime –imipenem antagonism test (CIAT),based on strong inducing effect of imipenem on these enzymes and the consequent antagonism with ceftazidime.

In this study , screening for inducible AmpC β -lactamase was done by the disk antagonism test, results showed that none of the isolates were positive for inducible AmpC β -lactamase .Similar finding was reported in a study conducted by Al-Hilli (2010) who found that among 7(28%) cefoxitin resistant isolates, no *K.pneumoniae* isolate was positive for inducible AmpC β -lactamase.

Results of the present study suggest that the encoded AmpC β - lactamases were rare among *K.pneumoniae* isolates collected from different clinical samples in Hilla hospitals, thus revealing the presence of only plasmid mediated AmpC β -lactamase.

5.15. Molecular Detection of AmpC β - Lactamase among Carbapenemase Positive *K. pneumoniae* Isolates

AmpC β -lactamases producing *Enterobacteriaceae* have become the major therapeutic challenge. The detection of AmpC -producing isolates is of significant clinical relevance since AmpC producers may appear susceptible to expanded –spectrum cephalosporins when initially tested (Pai *et al.*,2004; Thomson,2001 and Thomson,2010).This may lead to inappropriate antimicrobial regimens and therapeutic failure (Siu *et al.*,2003) .Currently, there are no Clinical Laboratory Standard Institute (CLSI) guidelines available for its optimal detection and confirmation (CLSI,2010).However ,phenotypic tests may be ambiguous and unreliable ,need careful interpretation .In addition ,the co-existence of ESBLs may mask its detection phenotypically(Mohamudha *et al.*,2012). Phenotypic tests do not differentiate between chromosomal AmpC genes and AmpC genes that are carried on plasmids, hence genotypic characterization is considered as the gold standard (Thomson,2001).

By PCR methods , *bla*AmpC genes was observed in 13/17(76.5%) isolates (Table 4-15 and Figure 4-18).All these isolates were cefoxitin resistant in cefoxitin susceptibility test (Figure 4-1). In one study Shahid *et al.* (2012) noted that among 16 carbapenem- resistant *E.coli* and *K.pneumoniae* , *bla*AmpC gene was detected in 14 (87.5%) isolates. Sobia *et al* .(2011) reported that 82.4% of *E.coli* and *K.pneumoniae* isolates harboring *bla*AmpC genes.

Genes encoding AmpC β-lactamases are commonly found on the chromosomes of several members of the family *Enterobacteriaceae* . The *ampC* is the structural gene for the AmpC enzyme. Recently , this gene has moved from the chromosome into self-transmissible plasmids. Gram-negative bacteria that lack an chromosomal AmpC enzyme (such as *K. pneumonia*e , *Salmonella* spp. and *P. mirabilis*) may acquire plasmids resulting in a stably depressed resistance phenotype. Obviously, the cefoxitin resistance and *bla*AmpC positive *K .pneumoniae* isolates may be acquired plasmidic *ampC* gene .Pai *et al.* (2004) mentioned that 50% of *K.pneumonia*e isolates received resistance to cefoxitin by the transmissibility plasmidic *bla*AmpC gene.

On the other hands, results demonstrated that the remaining 4 (23.5%) of cefoxitin resistant isolates were negative for *bla*AmpC gene (Figure 4-18). Cefoxitin resistance in AmpC non – producers could be due to some other resistance mechanisms. Hernandez-Alles *et al.* (2000) demonstrated that the interruption of a porin gene by insertion sequences is a common type of mutation that causes loss or decrease of outer membrane porin expression and increased cefoxitin resistance in *E. coli* and *Klebsiella* spp.

5.16. Emergence of Multiple *bla* –Genes Harboring *K.pneumoniae* Isolates

K.pneumoniae is one of the most common bacteria that cause nosocomial infections with multiple drug resistance because of its plasmid mediated β -lactamase (Jiang *et al.*,2012). Table (4-15) reports the finding of a plasmid harboring multiple resistance genes ,includings genes encoding carbapenemases ,ESBLs as well as AmpC β - lactamases. The results of this study showed that 76.5% of the carbapenemase harboring *K.pneumoniae* isolates carried ESBLs and AmpC β - lactamases.These results strongly suggest that there is co-existence of carbapenemase,ESBL and AmpC in these isolates. However ,in Taiwan similar results were recorded by Wu *et al.*(2011) who stated that among 82 ertapenem-resistant *K.pneumoniae* isolates ,the co-existence of ESBL and AmpC was found in 29 (35.4%) isolates.

Among 13 β - lactamase harboring *K.pneumoniae* , 5 isolates carried 7 genes, 6 isolates carried 8 genes and 2 isolates carried 9 genes. (Table 4-15). Present results suggest a clonal spread of multidrug resistant *K.pneumoniae* in Hilla hospitals. The co-existence of multiple *bla* genes in one isolates has been previously documented for *K.pneumoniae* ,with VIM-4,CTX-M and CMY-4 AmpC β - lactamase (Katri *et al.*, 2006); KPC, TEM, SHV and CTX-M β -lactamase (Carbal *et al* .,2012) ; NDM-1,TEM,SHV, CTX-M and AmpC β- lactamase (Bora and

Ahmed,2012).In Indian study , Shahid *et al.* (2012) pointed that among 16 carbapenem- resistant *E.coli* and *K.pneumoniae* ,2 isolates carried CTX-M, TEM and VIM whereas one isolate carried CTX-M,TEM, SHV and VIM genes.

The finding that 13/17 (76.5%) isolates were carried at least 7 of the β - lactamase genes is worrying. Ktari *et al.*(2006) stated that the simultaneous production of three β - lactamases by *K.pneumoniae* deserves to be highlighted. The accumulation of resistance genes in *K.pneumoniae* isolates observed in this study ,imposes limitations in the therapeutic options available for the treatment of infections caused by this pathogen in Hilla. These results should alert to establish rigorous methods for detection and thus more efficiently control the dissemination of antibiotic resistance genes in the hospital environment. Moreover, measures to control excessive use of antimicrobials in hospitals must also be taken.

5.17. Transfer of Carbapenem Resistance

The conjugation process is the transfer of DNA element directly from one bacterial cell to another by a mechanism that requires cell- to - cell contact. It has been considered as a major pathway for horizontal gene transfer among bacteria (Normark and Normark,2002 and Somkiat *et al.*,2007). One purpose of this study was to investigate the mobility of carbapenemase genes by transconjugation. However, conjugation experiments and PCR assay demonstrated that carbapenemase positive *K.pneumoniae* isolates (K13 and K14) were able to transfer *bla*VIM gene in transconjugants (Table 4-16 and Figure 4-19). In the present investigation ,conjugation experiments proved that *bla*VIM gene was transferable . In Italian study , a transferable plasmid encoding the VIM-

4 metallo β -lactamase was detected in isolates of *K.pneumoniae* and *E. cloacae* obtained from a single patient under carbapenem therapy (Luzzaro *et al.*,2004). Success in transferring the VIM gene by transconjugation has been frequently reported (Tato *et al.*,2007 and Sanchez-Romero *et al.*,2011), and this perhaps contributed to the rapid dissemination of resistance to carbapenems among bacteria in Hilla hospitals.

On the other hand *bla*NDM-1 and *bla*OXA-23 presented in *K.pneumoniae* K13 and K14 can not transferred by conjugation. The failure in transferring *bla*NDM-1 and *bla*OXA-23 indicates that these genes may be located on non–transferable plasmid. Similar findings also was recorded by El-Herte *et al.* (2012) who stated that all *K.pneumoniae* isolates were unable to transfer *bla*NDM-1 by conjugation experiments. *Bla*OXA-48 has also been detected on non –transconjugative plasmids (Li *et al.*,2012).

The conjugation frequency for these transconjugants ranged from 4.6×10^{-2} in *K.pneumoniae* K13 to 6.5×10^{-2} in *K.pneumoniae* K14 (Table 4-17). In another study conducted by Loi *et al* (2006) reported that VIM-1 was readily transferred by conjugation from all VIM-1 producing *K.pneumoniae* to *E.coli* at high frequency (10^{-4} to 10^{-5}). Other study characterized that *bla*VIM-1 gene was successfully transferred from all Gram –negative isolates to *E.coli*26R793 with frequencies ranging from 1.5×10^{-1} to 1.79×10^{-2} (Sianou *et al.*,2011).

Antibiotic susceptibilities for the two transconjugants ,TCK13 and TCK14 showed much higher MICs values (>32 µg /ml) for imipenem and meropenem relative to those of the recipient (Table 4-16 and Figure 4-20).The present results suggested that the higher resistance to imipenem and meropenem among transconjugants as their donor isolates might imply that the VIM gene was transferred and expressed.

PCR assay confirmed that the *K.pneumoniae* isolates K13 and K14 were able to transfer the *bla*SHV, *bla*CTX-M and *bla*AmpC in transconjugants (Table 4-16; Figures 4-21 ,4-22 and 4-23). Co–transfer of *bla*VIM with other resistant determinants such as *bla*CTX-M, *bla*SHV and *bla*TEM-1 were detected in one conjugation experiments (Samuelsen *et al* .,2011).In another study, Tokatlidou *et al.* (2008) found that *bla*VIM and *bla*CMY genes were resided on the same transferable plasmid. The presence of VIM and other *bla* genes on same plasmid is one of several possible explanation for this association.

The association of *bla*VIM and other *bla* genes severely limit the therapeutic options for the treatment infection with VIM-producing organisms. It seems necessary to measures for prevention of their dissemination and the spread the genetic elements harboring the metallo β- lactamase encoding genes. In the absence of such measures,MBL-producing *K.pneumoniae* may be significantly contribute to rapid emergence of nosocomial infections that antibiotic can not treat.

Conclusions

1- Class B carbapenemase seems to be a predominant cause of carbapenem resistance among clinical isolates of *K.pneumoniae* in Hilla hospitals.

2- The vast majority of carbapenemase producing *K.pneumoniae* isolates were found to be associated with ESBL and AmpC β - lactamase genes, the most predominate among these were *bla*TEM, *bla*SHV, *bla*CTX-M and *bla*OXA-1 .

3- The co-existence of multiple plasmid genes (carbapenemase, ESBL and AmpC) in *K.pneumoniae* isolates is frequently accompanied with extensive-drug and pandrug resistance.

4- Emergence of pandrug resistant isolates threatening the antimicrobial treating program in Hilla hospitals.

5- Successful co-transmission of VIM and other *bla*-genes by conjugation may contribute to the alarming rates of carbapenem resistant *K.pneumoniae* isolates in Hilla hospitals.

Recommendations

1-Phenotypic and molecular methods (PCR) identify the presence of carbapenemase producing isolates should be applied routinely in Hilla hospitals.

2-Continued surveillance of carbapenem resistance isolates and understanding other molecular mechanisms of resistance and disseminations play a pivotal role in controlling spread and guiding antimicrobial therapy.

3-Genetics relatedness among carbapenem -resistant *K.pneumoniae* isolates by pulsed field gel electrophoresis need further studies.

4-DNA sequencing of NDM-1 gene to confirm that the precise gene was amplified.

5-RT PCR to study gene expression need further investigations.

References

Abic,M; Hujer,A.M. and Bonomo,R.A.(2006).What's new in antibiotic resistance?Focus on beta-lactamases. Drug.Resist.Updat.9:142-156

Acheampong, D.O.; Boamponsem, L. K. and Feglo, P.K. (2011). Occurrence and species distribution of *Klebsiella* isolates : a case study at komfo Anokye teaching hospital (Kath) in Ghana. Adv. Appl. Sci .Res. 2(40) :187-193.

Aktas, Z.; Satana, D.; Kayacan, C.; Ozbek, B.; Gurler, N.; Somer, A.; Salman, N. and Aydin, A. E. (2012). Carbapenem resistance in Turkey: repeat report on OXA-48 in *Klebsiella pneumoniae* and first report on IMP-1 beta- lactamsae in *Escherichia coli*. Afr. J. Microbiol. Res. 6(17): 3874-3878.

Al-Asady , F.M.H.(2009) .Bacteriological study on extended-spectrum beta- lactamases produced by *Enterobacteriaceae* isolated from children with bacteremia in Hilla city .M.Sc. Thesis .College of Medicine,Babylon University.

Al- Charrakh, A. H. (2005). Bacteriological and genetic study on extended- spectrum β - lactamases and bacteriocins of *Klebsiella* isolated from Hilla city. Ph. D. Thesis. College of Sciences, Baghdad University.

Al- Hilli, Z. B. (2010). Dissemination of β - lactamases in *Escherichia coli* and *Klebsiella* spp. isolated from Merjan teaching hospital in Hilla city. M. Sc. Thesis. College of Sciences, Kufa University.

AL- Jasser, A. M. (2006). Extended- spectrum beta- lactamases (ESBLs): a global problem. Kuw. Med. J. 38(3): 171-185.

AL- Muhannak, F. H. N. (2010). Spread of some extended spectrum beta-lactamases in clinical isolates of Gram-negative bacilli in Najaf. M. Sc. Thesis. College of Medicine. Kufa University.

Al-Saedi, I.A.B. (2000).Isolation and identification of *Klebsiella pneumoniae* from various infection in Hilla Province and detection of some virulence factors associated in their pathogenicity .M.Sc.Thesis.College of Sciences ,Babylon University.

Al-Sehlawi,Z.S.R.(2012).Occurrence and characterization of AmpC β - lactamases in *Klebsiella pneumoniae* from some medical centers in Najaf .Ph.D.Thesis. College of Sciences ,Babylon University.

AL- Shara, M. A. (2011). Emerging antimicrobial resistance of *Klebsiella pneumoniae* strains isolated from pediatric patients in Jordan. N. Iraqi. J. Med. 7(2): 29-32.

Alegria, C. R. de.; Rodriguez- Bano, J.; Cano, M. E.; Hernandez- Bello, J. R.; Calvo, J.; Roman, E.; Diaz, M.A.; Pascual, A. and Martinez- Martinez, L. (2011). *Klebsiella pneumoniae* strains producing extended- spectrum β - lactamases in Spain: microbiological and clinical features. J. Clin. Microbiol. 49(3): 1134-1136.

Ali, S. Q.; Zehra, A.; Naqvi, B. S.; Shah, S, and Bushra, R. (2010). Resistance pattern of ciprofloxacin against different pathogens. Oman. Med. J. 25: 294-298.

Allen, B. L.; Gerlach, G. F. and Clegg, S. (1991). Nucleotide sequence and functions of mrk determinants necessary for expression of type 3 fimbriae in *Klebsiella pneumoniae*. J. Bacteriol. 173: 916- 920.

Alves, M. S.; Dias, R. C.; de Castro, A. C.; Riley, L. W. and Moreira, B. M. (2006). Identification of clinical isolates of indole-positive and indole-negative *Klebsiella* spp. J .Clin. Microbiol. 44: 3640- 3646.

Amador, P.R.F .; Prudencio ,C. and Brito, L.(2009). Resistance to β- Lactams in bacteria isolated from different types of portuguese cheese .Int .J. Mol .Sci. 10: 1538- 1551.

Ambler, R. P. (1980). The structure of β-lactamases. Philos Trans R Soc Lond B Biol .Sci .289: 321-331.

Ambler, R. P.; Coulson, A. F.; Frere, J. M.; Ghuysen, J. M.; Joris, B.; Forsman, M.; Levesque, B. C.; Tiraby, G. and Waley, S. G. (1991). A standard numbering scheme for the class A β - lactamases. Biochem. J. 276: 269-272.

Amin, A.; Ghumro, P.B.; Hussain, S. and Hameed, A.(2009). Prevalence of antibiotic resistance among clinical isolates of *Klebsiella pneumoniae* isolated from a tertiary care hospital in Pakistan. Mal.J.Microbiol. 5 (2): 81-86.

Aminizadeh, Z. and Kashi, M.S. (2011). Prevalence of multi-drug resistance and pandrug resistance among multiple Gram-negative species: experience in one teaching hospital, Tehran, Iran. Int. Res. J. Microbiol. 2(3): 090-095.

Amjad,A. ; Mirza ,I.A.;Abbasi ,S.A.; Farwa,U;Malik,N. and Zia, F. (2011).Modified Hodge test:a simple and effective test for detection of carbapenemase production .Iran J. Microbiol. 3(4):189-193.

Amyes, S.G. (2003). Resistance to beta-lactams-the permutations.J. Chemother. 15 (6): 525-535.

nderson,K.F.;Lonsway,D.R.;Rasheed,J.K.;Bidlle,J.;Jensen,B.McDougal , L.K.; Carey, R.B.; Thompson ,A.; Stocker, S.; Limbago , B. and Patel,J.B.(2007).Evalution of methods to identify the *Klebsiella pneumoniae* carbapenemase in *Enterobacteriaceae*. J.Clin.Microbiol.4 5(8)2723-2725.

Andes,D.R. and Craig,W.A.(2005).Cephalosporins in Principle and Practice of Infectious Diseases,eds,Mandell,G.L.;Bennett,J.E.and Dolin,R.Philadelphia ,PA.:Churichill Livingstone .294-311.

Andrews, J. M. (2001). Determination of minimum inhibitory concentrations. J. Antimicrob. Chemother. 48: suppl. S1, 5-16.

Arakawa, Y.; Shibata, N.; Shibayama,K.; Kurokawa, H.; Yagi, T; Fujiwara, H. and Goto,M.(2000). Convenient test for screening metallo-ß-lactamase-producing Gram negative bacteria by using thiol compounds. J. Clin. Microbiol. 38:40-43.

Aroutcheva, A.; Gariti, D.; Simon, M.; Shott, S.; Faro, J.; Simoes, A.;Dominique,A.;Melissa,G.and Gurquis,A. (2001) . Defence factor of vaginal Lactobacilli. Am. J. Obstet. Cynecol. 185: 375-379.

Arnold,R.S.;Thom,K.A.;Sharma,S.;Phillips,M.;Johnson,J K.and Morgan, D.J .(2011).Emergence of *Klebsiella pneumoniae* carbapenemase (KPC)-producing bacteria .South. Med. J.104(1):40-45.

Askarian, M.; Reza, S. H.; Parastoo K. and Ojan, A. (2004). Incidence and outcome of nosocomial infections in female burn patients in Shiraz, Iran. Am. J. of Infect. Cont. 32:23-26.

Azimi , L .; Motevallian , A; Namvar , A.E. ; Asghari, B. and Lari, A.R. (2011). Nosocomial infections in burned patients in Motahari hospital ,Tehran ,Iran .Derm. Res. and Prac. 1-4.

Babini,G.S.and Livermore,D.M.(2000).Are SHV β - lactamse universal in *Klebsiella pneumoniae*?J.Antimicrob.Agent.Chemother. 44:2230.

Bae, I. K.; Jang. S.J.; Kim, J.; Jeong, S. H.; Cho, B. and Lee, K. (2011). Interspecies dissemination of the *bla* gene encoding PER-1 extended- spectrum β - lactamase. Antimicrob. Agents. Chemother. 55(3): 1305-1307.

Bagley, S.; Seidler, R. J. and Brenner, D. J. (1981). *Klebsiella planticola* sp. nov.: a new species of *Enterobacteriaceae* found primarily in nonclinical environments. Curr .Microbiol. 6: 105-109 .

Bali, E. B.; Acik, L. and Sultan, N. (2010). Phenotypic and molecular characterization of SHV, TEM, CTX-M and extended – spectrum β - lactamase produced by *Escherichia coli*, *Acinetobacter baumanii* and *Klebsiella* isolates in a Turkish hospital. Afr. J. Microbiol. Res. 4(8) : 650- 654.

Baquero F., Negri MC, Morosini Ml and Blazquez J.(1997). The antibiotic selective process: Concentration- specific amplification of low- level resistant populations. Ciba. Found. Symp .207:93-111.

Baraniak,A.; Fiett,J.; Sulikowska,A.; Hryniewicz, W.and Gniadkowski, .(2002).Countrywide spread of CTX-M-3 extended spectrum beta lactamase producing microorganisms of the family *Enterobacteriaceae* in Poland .Antimicrob. Agents. Chemother.46:151-159.

Baron, E. J. and Finegold , S. M. (1994). Baily and Scott's: Diagnostic Microbiology. 8th ed. Mosby company, Missouri.

Barros , J.C.S.; Bozza , M.; Gueiros- Filho, F.J; Bello, A.B.; Lopes, U.G and Pereira , J.A.A. (1999). Evidences of gentamicin resistance amplification in *Klebsiella pneumoniae* isolated from faeces of hospitalized newborns.Mem . Inst. Oswaldo Cruz. Rio de Janeiro. 94(6): 795-802.

Bartges, J. (2007) .Evaluation of an alternative method for dissolving infection induced straiten stones .urinary studies unit .Veterinary college research .University of Tennessee.

Basavaraj,C.; Jyothi, P.and Basavaraj,P.V.(2011).The prevalence of ESBL among *Enterobacteriaceae* in a tertiary care hospitals of North Karnataka,India.J.Clin.Dia.Res.5(3):470-475.

Bauer, A. W.; Kirby, W.M.M.; Sherris, J.C and Track, M. (1966). Antibiotic susceptibility testing by standardized single disc method .Am.J. Clin.Pathol.45: 493-496.

Bauernfeind, A.; Grim, H.; and Schweighart, S. (1990). A new plasmidic cefotaximase in a clinical isolate of *Escherichia coli*. Infection. 18: 294-298.

Beg,Q.Z.; Al-Hazimi,A.M.; Ahmed,M.Q.; Fazaludeen,M.F. and Shabeen, R.(2011).Resistant bacteria a threat to antibiotics. J. Chem. Pharm.Res. 3(6):715-724.

Ben-David,D. ;Kordevani,R. ;Keller,N. ;Tal,I .;Marzel,A. ;Gal-Mor,O. and Rahav,G. (2012). Outcome of carbapenem resistant *Klebsiella pneumoniae* bloodstream infections. Clin. Microbiol. Infect.18:54-60.

Ben-Hamouda, T.; Foulon, T.; Ben-Cheikh-Masmoudi, A.; Fendri, C.; Belhadj, O. and Ben-Mahrez, K. (2003). Molecular epidemiology of an outbreak of multiresistant *Klebsiella pneumoniae* in a Tunisian neonatal ward. J. Med. Microbiol. 52: 427-433.

Benson, H.J.(1998). Microbiological applications, Lab manual in general microbiology, 7th edition, McGraw-Hill companies.

Birgy, A.; Doit, C.; Mariani- Kurkdjian; P.; Genel, N.; Faye, A.; Arlet, G. and Bingen, E. (2011). Early detection of colonization by VIM-1 producing *Klebsiella pneumoniae* and NDM-1 producing *Escherichia coli* in two children returning to France. J. Clin.Microbiol. 49(8): 3085-3087.

Bogaerts, P.; Montesinos, I.; Villa labos, H.; Blairon , L.; Deplano, A. and Glupczy.(2010). *Klebsiella pneumoniae* isolates of sequence type 58 producing KPC-2 carbapenemase in Belgium . I. Antimicrob. Chemother. 65: 361-376.

Bonnet, R. (2004). Growing group of extended spectrum β - lactamases: the CTX-M enzymes. Antimicrob. Agents Chemother. 48 (1):1-14.

Bonomo,R.A.(2011).New Delhi-metallo-β-lactamase and multidrug resistanc:a global SOS?.Clin.Infect.Dis.52(4)485-48.

Bora, A. and Ahmed, G. (2012). Detection of NDM-1 in clinical isolates of *Klebsiella pneumoniae* from North East India. J. Clin. Diag. Res. 6(5): 794-800.

Bouchakour, M.; Zerouali, K.; Glaude, J.D.P. G.; Amarouch ,H. ; Mdaghri, N.E. ; Courvalin , P. and Timinouni, M. (2010). Plasmid mediated quinolone resistance in expanded spectrum beta- lactamase producing *Enterobacteriaceae* in Morocco. J.Infect. Dev. Ctries. 4(12): 799- 803.

Boucher ,H.W; Talbot, G.H.; Bradley, J.S;Edwards, J.E.;Gilbert, D.; Rice,L.B.;Scheld,M.;Spellberg, B. and Bartlett,J. (2009). Bad bugs, no drugs :no ESKAPE! an update from the Infectious Diseases Society of America . Clin. Infect. Dis. 48 :1 -12.

Bowler, P.G.; Duerden, B.I. and Armstrong, D.G. (2001). Wound microbiology and associated approaches to wound management. Clin. Microbiol. Res. 14: 244-269.

Bradford, P. A. (2001). Extended spectrum β - lactamases in the 21th century: characterization, epidemiology, and detection this important resistance threat. Clin. Microbiol. Rev. 14: 933-951.

Braiteh, F. and Golden, M. P. (2007). Cryptogenic invasive *Klebsiella pneumoniae* liver abscess syndrome. Int. J. Infect. Dis. 11: 16-22.

Brakouv,M.P. ;Eber,M.R.; Klein,E.Y.; Mogan, D.J.and Laxminarayan,R. (2013). Trends in resistance to carbapenem and third generation cephalosporins among clinical isolates of *Klebsiella pneumoniae* in the United States,1999-2010.Infect .Control.Hospit. Epidemiol. 34(3):1-9

Branger , C. ; Lesimple , A.L. ; Bruneau , B.; Berry ,P. and Zechovsky , N. (1998). Long –term investigation of the clonal dissemination of *Klebsiella pneumoniae* isolates producing extended- spectrum β- lactamases in a university hospital .J .Med. Microbiol .47:201-209.

Bratu, S.; Brooks, S. and Burney, S. (2007). Detection and spread of *Escherichia coli* possessing the plasmid- borne carbapenemase KPC-2 in Brooklyn, New York. Clin. Infect. Dis. 44: 972-973.

Bratu, S., Landman,D; Alam,M.; Tolentino,E. and Quale,J. (2005b). Detection of KPC carbapenem-hydorlyzing enzymes in *Enterobacter* spp. from Brooklyn, New York. Antimicrob. Agents Chemother. 49:776-778.

Bratu, S.; Landman , D.; Haag , R. (2005a). Rapid spread of carbapenem - resistant *Klebsiella pneumoniae* in New York City: a new threat to our antibiotic armamentarium. Arch. Intern. Med. 165: 1430-1435.

Brisse, S.; Grimont, F. and Grimont, P. A. D. (2006). The Genus *Klebsiella*. Prokaryotes. 6: 159-196.

Brooks , G.F.; Butel, J.S. and Morse, S.A. (2001) .Enteric Gram- negative rods (*Enterobacteriaceace*) .In :Brooks, G.F; Butel J.S.; Morse , S.A. :Jawetz –Melnick, and Adelberg's Medical Microbiology. 22nd ed. McGraw-Hill .U SA.

Brown, S. and Amyes, S. (2006). OXA β - lactamases in *Acinetobacter*: the story so far. J. Antimicrob. Chemother. 57: 1-3.

Bulik, C.C. and Nicolau, D.P. (2011). Double- carbapenem therapy for carbapenemase- producing *Klebsiella pneumoniae*. Antimicrob. Agents Chemother. 55(6): 3002- 3004.

Bush, K. (1989a). Classification of β-lactamases: groups 1, 2a, 2b, and 2b. Antimicrob .Agents Chemother. 33: 264-270.

Bush, K. (1989b). Classification of β-lactamases: groups 2c, 2d, 2e, 3, and 4.Antimicrob .Agents Chemother. 33: 271-276.

Bush, K. (1989c). Characterization of β-lactamases. Antimicrob. Agents Chemother. 33: 259-263.

Bush, K. (1996). Other β-lactams. Antibiotic and Chemotheraphy ́Grady,F.O.; Lambert, H.P.; Finch,R.G. and Greenwood,D. New York, Churchill. Livingstone PP: 306-327.

Bush, K. (2010). Alarming β-lactamase –mediated resistance in multidrug-resistant *Enterobacteriaceae*.Curr.Opin.Microbiol.13 (5):558-564.

Bush, K. and Jacoby, G. A. (2010). Updated functional classification of β - lactamases. Antimicrob. Agents Chemother. 54(3): 969-976.

Bush, K.; Jacoby, G. and Medeiros, A. (1995). A functional classification scheme for β - lactamases and its correlation with molecular structure. J. Antimicrob. Agents Chemother. 39: 1211-1233.

Bush, K.; Macalintal, C.; Rasmussen, B.A.; Lee, V.J. andYang, Y.(1993). Kinetic interactions of tazobactam with beta- lactamases from all major structural classes. Antimicrob. Agents Chemother. 37: 851-858.

Cabral, A.B.; Melo, R.C.A. ;Maciel , M.A.V. and Lopes, A.C.S. (2012). Multidrug resistance genes, including *bla*KPC and *bla*CTX-M-2 among *Klebsiella pneumoniae* isolated in Recife, Brazil.Rev. Bras. Med.Trop. 45(5) :572-578.

Cai, J. C.; Zhou, H. W.; Zhang, R. and Chen, G-X. (2008). Emergence of *Serratia marcescens, Klebsiella pneumoniae,* and *Escherichia coli* isolates possessing the plamid- mediated carbapenem-hydrolyzing β - lactamase. Antimicrob. Agents Chemother. 52 (6): 2014-2018.

Cai, X- F.; Sun, J- M.; Baol;L- S. and Li, W-b. (2011). Distribution and antibiotic resistance of pathogens isolated from ventilator-associated pneumonia patients in pediatric intensive care unit. World. J. Emerg. Med.2(2):117-121.

Cao,V.T.B.; Arlet,G.; Ericsson,B-M.; Tammelin,A.; Courvalin,P. and Lambert,T. (2000).Emergence of imipenem resistance in *Klebsiella pneumoniae* owing to combination of plasmid mediated CMY-4 and permeability alteration. J.Antimicrob. Chemother.46:895-900.

Cambray,G.; Guerout, A-M. and Mazel,D. (2010). Integrons. Annu.Rev.Genet.44:141-166.

Cantarelli, V.V.; Inamine, E.; Brodt, T. C. Z.; Secchi, C.; Cavalcante, B. C.and Pereira, F. de. S. (2007). Utility of the ceftazidime-imipenem antagonism test (CIAT) to detect and confirm the presence of inducible AmpC beta- lactamases among *Enterobacteriaceae*. Braz. J. Infect. Dis. 11(2):237-239.

Carattoli, A. (2009). Resistance plasmid families in *Enterobacteriaceae*. Antimicrob. Agents Chemother. 53:2227-2238.

Carter, J. S; Bowden, F. J.; Bastian, I.; Myers, G. M.; Sriprakash, K. S.and Kemp, D. J. (1999). Phylogenetic evidence for reclassification of *Calymmatobacterium granulomatis* as *Klebsiella granulomatis* comb. nov. Int. J. Syst. Bacteriol. 49: 1695-1700.

Chart, H. (2007). *Klebsiella* , *Enterobacter*, *Proteus* and other enterobacteria, In: Greenwood, D.; Slack, R.; Peutheror, J. and Barer, M. (eds). Medical Microbiology. A Guide to Microbial Infections: Pathogenesis, Immunity, Laboratory Diagnosis and Control. 17th ed. Churchill livingstone.

Chen, L.F.; Anderson, D.J. and Paterson, D.L. (2012). Overview of the epidemiology and the threat of *Klebsiella pneumoniae* carbapenemases (KPC) resistance. Infect. Drug. Resist. 5: 133-141.

Chia, J- H.; Su, L-H.; Lee, M-H.; Kuo, A-J.; Shih, N-Y.; Siu, L. K. and Wu, T-L (2010). Development of high – level carbapenem resistance in *Klebsiella pneumoniae* among patients with prolonged hospitalization and carbapenem exposure. Microbiol. Drug. Resist. 16(4): 317-325.

Chiangjong, W. (2006) .Study of extended- Spectrum beta- lactamase (ESBL) producing *Klebsiella pneumoniae* genotypic and phenotypic characteristics .M.SC .Thesis .Faculty of Graduate Studies .Mahidol University.

Chihara, S.; Okuzumi, K.; Yamamoto, Y.; Oikawa, S. and Hishinuma, A. (2011). First case of New Delhi metallo- beta- lactamase 1-producing *Escherichia coli* infection in Japan. Clin. Infect. Dis. 52:153-154.

Clinical and Laboratory Standards Institute (CLSI). (2010). Performance standards for antimicrobial susceptibility testing; 20th. Informational Supplement. Approved standard M07-A8. Clinical and Laboratory Standards Institute.

Collee, J. G.; Fraser, A. G.; Marmiom, B. P.; and Simmon, A. (1996). Mackie and McCarteny Practical Medical Microbiology. 4th ed. Churchill Livingstone Inc., USA.

Coudron, P. E.; Hanson, N. D. and Climo, M. W. (2003). Occurrence of extended spectrum and AmpC β - lactamases in bloodstream isolates of *Klebsiella pneumouiae* isolates harbor plasmid-mediated FOX-5 and ACT-1 AmpC β - lactamases. J. Clin. Microbiol. 41: 772-777.

Coudron, P. E.; Moland, E. S.; and Sanders, C. C.(1997).Occurence and detection of extended -spectrum β-lactamases in members of the family *Enterobacteriacea* at a Veterans Medical Center :seek and you may find. J. Clin. Microbiol. 35: 2593-2597.

Coudron, P.E.; Molond, E. S. and Thomson, K. S. (2000). Occurrence and detection of AmpC beta- lactamases among *Escherichia coli* , *Klebsiella pneumoniae*, and *Proteus mirabilis* isolates at a Veterans medical center. J. Clin. Microbiol. 38: 1791-1796.

Courvalin, P. (2006). Antibiotic resistance: the pros and cons of probiotics. Digestive and Liver Dis. 38 (2): 261-265.

Cross, A.; Allen, J. R.; Burke, J.; Ducel, G.; Harries , A.; John, J.; Johnason, D.; Lew , M.; McMillan, B.; Meers, P.; Skalova , R.; Wenzel, R. and Jenney, J. (1983). Nosocomial infections due to *Pseudomonas aeroginosa*. Review of recent trends. Rev. Infect. Dis. 5 (Suppl.): 837- 845.

Dalla- Costa , L. M.; Coelho , J. M.; Souza, H. A. ; Castro, M. E. S.; Stier, C. J. N.; Bragagnolo , K.L.; Rea- Neto- A.; Penteado-Filho, S. R.; Livermore, D. M. and Woodford , N. (2003). Outbreak of carbapemem – resistant *Acinetobacter baumannii* producing the OXA- 23 enzyme in Curitiba, Brazil. J. Clin. Microbiol. 41:3403-3406.

Datta, N. and Kontomichalou, P. (1965). Penicillinase synthesis controlled by infectious R factors in *Enterobacteriaceae*. Nature 208: 239-241.

Decre , D. ; Gachot , B.; lucet , J. C. ; Arlet ,G.;Byergogne-Berezin,E.and Regnier,B. (1998) . Clinical and bacteriologic epidemiology of extended – spectrum β- lactamase –producing strains of *Klebsiella pneumoniae* in a medical intensive care unit .Clin . Infect . Dis. 27:843- 844.

Deshmukh, D. G.; Damle, A. S.; Bajaj, J. K. and Bhakre, J. B. (2011). Metallo β - lactamase producing clinical isolates from patients of a teriary care hospital. J. Lab. Physicians. 3 (2). 93-97.

Deshpande,P.;Rodrigues,C.:Shetty,A.;Kapadia,F.;Hedege,A.and Soman ,R.(2010). New-Delhi metallo beta -lactamase (NDM-1) in *Enterobacteriaceae*: the treatment options with the carbapenems are comprimized .J.Asso. Physians.India.58:147-149.

Drancourt, M.; Bollet, C.; Carta, A. and Rousselier, P. (2001). Phylogenetic analyses of *Klebsiella* species delineate *Klebsiella* and *Raoultella* gen. nov., with 201 description of *R. ornithinolytica* comb. nov., *R. terrigena* comb. nov. and *R. planticola* comb. nov. Int. J .Syst. Evol. Microbiol .51: 925-932.

Drlica, K. and Zhao, X. (1997). DNA gyrase, topoisomerase IV, and the 4-quinolones. Microbiol .Mol. Biol .Rev. 61: 377-392.

Dugal ,S. and Fernandes ,A. (2011) Carbapenem hydrolyzing metallo-β - lactamases :A. Review .Int .J.Curr. Pharm. Res .3 :9-16.

Dzidic,S.; Suskovic,J.and Kos,B. (2008). Antibiotic resistance mechanisms in bacteria: biochemical and genetics aspects.Food Technol. Biotechnol. 46:11-21.

Ejikeuqwu, P. C.; Ugwu, C. M.; Araka, C. O.; Gugu, T. H.; Iroha, I. R.; Adikwu, M. U. and Esimone, C. O. (2012). Imipenem and meropenem resistance amongst ESBL producing *Escherichia coli* and *Klebsiella pneumoniae* clinical isolates. Int. Res. J. Microbiol. 3(10): 339-344.

EL- Dien, A. and Kashif, M. T. (2012). A study on occurrence of plasmid AmpC β - lactamase among Gram- negative clinical isolates and evaluation of different methods used for their detection. J.Appl. Sct. Res.8(4):2280– 2285.

El- Herte, R. I.; Araj, G. F.; Matar, G. M.; Baroud , M.; Kanafani , Z. A. and Kanj,S.S.(2012). Detection of carbapenem-resistant *Escherichia coli* and *Klebsiella pneumoniae* producing NDM-1 in Lebanon. J. Infect. Dev. Ctries. 6(5):457-461.

Elufisan, T. O.; Oyedara, O. O. and Oyelade ,B. (2012). Updates on microbial resistance to drugs. Afr. J. Microbiol. Res. 6(23): 4833-4844.

Erac,B.and Gulay,Z.(2007).Molecular epidemiology of PER-1 extended spectrum β - lactamase among Gram-negative bacteria isolated at a tertiary care hospital.Floia.Microbiol.52(2);535-541.

Falagas, M.E.; Bliziotis, I.A.; Kasiakou, S.K. ; Samonis G.; Athanassopoulou, P. and Michalopoulos, A.(2005). Outcome of infections due to pandrug –resistant (PDR) Gram- negative bacteria. BMC. Infect .Dis.5: 24

Falagas, M. E.; Rafailidis, P. I.; Kofteridis, D.; Virtzili, S.; Chelvatzoglou, F. C.; Papaioannon, V.; Maraki, S.; Samonis, G. and Michalopoulos, A. (2007). Risk factors of carbapenem-resistant *Klebsiella pnenmoniae* infections: a matched case-control study. J. Antimicrobiol. Chemother. 60: 1124- 1130.

Falagas, M.E. ; Rafailidis ,P.I; Mathaiou, D.K.; Virtzili, S. ; Nikita,D. and Michalopoulos, A.(2008). Pandrug –resistant *Klebsiella pneumoniae* ,*Pseudomonas aeruginosa* and *Acintobacter baumannii* infections : characteristics and outcome in a series of 28 patients. Int.J.Antimicrob.Agents. 32(5) :450 - 454.

Fallah,F.;Taherpour,A.;Vola,M.H. and Hashemi,A.(2011).Global spread of New Delhi metallo-beta-lactamase-1(NDM-1). Iran. J.Clin. Infect. Dis.6(4):171-176.

Fam, N.S. and El-Damarawy, M.M. (2008). CTX-M-15 extended-spectrum beta-lactamases detected from intensive care unit of an Egyptian medical research institute. Res. J .Med. and Medical Sciences. 3 (1): 84-91.

Feldman, C., Ross, S., Mahomed, A. G., Omar, J. and Smith, C. (1995). The aetiology of severe community-acquired pneumonia and its impact on initial, empiric, antimicrobial chemotherapy. Respir .Med. 89: 187-192.

Ferragut, C.; Izard, D.; Gavini, F.; Kersters, K.; De Ley, J. and Leclerc, H. (1983). *Klebsiella trevisanii* : a new species from water and soil. Int .J .Syst. Bacteriol .33: 133-142.

Ferreira, C. M.; Ferreira, W. A.; Almeida, N. C. O.; Naveca, F. G. and Barbosa, M. d. G. (2011). Extended- spectrum beta- lactamase-producing bacteria isolated from hematologic patients in Manaus, state of Amazonas, Brazil. Brazil. J. Microbiol. 42:1076-1084.

Flonta, M.; Lupse, M.; Craciunas, C.; Almas, A. and Carstina, D. (2011). Ertapenem resistance among extended- spectrum- β - lactamase producing *Klebsiella pneumoniae* isolates. Therapeu. Pharmacol. Clin. Toxicol. 15(2): 121-125.

Forssten, S. (2009). Genetic basis and diagnostics of extended- spectrum β - lactamases among *Enterobacteriaceae* in Finland. M. Sc. Thesis. University of Turku.

Ford, P. J. and Avison, M. B. (2004). Evolutionary mapping of the SHV β-lactamase and evidence for two separate IS*26*-dependent *bla*$_{SHV}$ mobilization events from the *Klebsiella pneumoniae* chromosome. J. Antimicrob . Chemother .54: 69-75.

Galimand, M.; Sabtcheva, S.; Courvalin, P. and Lambert, T. (2005). Worldwide disseminated *arm*A aminoglycoside resistance methylase gene is borne by composite transposon Tn 1548. Antimicrob. Agents Chemother. 49: 2949-2953.

Garcia- Fernandez, A.; Miriagou, V.; Papagiannitsis; Giordano, A.; Venditti, M.; Mancini, C. and Carattoli, A. (2010). An ertapenem resistant extended-spectrum β - lactamase-producing *Klebsiella pneumoniae* clone carries a novel OmpK36 porin variant. Antimicrob. Agents Chemother. 54(10): 4178-4184.

Gasink, L. B; .Edelstein, P.H.; Lautenbach, E.; Synnestvedt, M. and Fishman,N.O. (2009). Risk factor and clinical impact of *Klebsiella pneumoniae* carbapenemase producing *K.pneumoniae*. Infect. Control. Hosp. Epidemiol.30:1180-1185.

Gavini, F.; Izard, D.; Grimont, P. A. D.; Beji, A.; Ageron, E. and Leclerc, H. (1986). Priority of *Klebsiella planticola* Bagley, Seidler, and Brenner 1982 over *Klebsiella trevisanii* Ferragut, Izard, Gavini, Kersters, DeLey, and Leclerc 1983. Int. J. Syst. Bacteriol .36:486-488.

Ghafourian, S., Sekawi, Z. b.; Sadeghifard, N., Mohebi, R.; Neela, V. K.; Maleki, A.; Hematian, A.; Rahbar, M.; Raftari, M. and Ranjbar, R. (2011). The prevalence of ESBLs producing *Klebsiella pneumoniae* isolates in some major hospitals, Iran. The. Open. Microbiol. J. 5: 91-95.

Ghuysen, J. M. (1988). Bacterial active-site serine penicillin-interactive proteins and domains: mechanism, structure, and evolution. Rev. Infect. Dis .10: 726-732.

Giakkoupi, P.; Pappa, O.; Polemis , M.; Vatopoulos, A. C.; Miriagou, V.; Zioga, A.; Papagiannitsis, C. C. and Tzouvelekis , L. S. (2009). Emerging *Klebsiella pneumoniae* isolates coproducing KPC-2 and VIM-1 carbapenemases. Antimicrob. Agents Chemother. 53(9): 4048-4050.

Giakkoupi, P.; Tzouvelekis, L. S.; Tsakris, A.; Loukova, V.; Sofianou, D. and Tzelepi, E. (2000). IBC-1, a novel integron-associated class A β-lactamase with extended-spectrum properties produced by an *Enterobacter cloacae* clinical strain. Antimicrob .Agents Chemother. 44: 2247-2253.

Giedraitiene, A.; Vitkauskiene, A.; Naginiene, R. and Pavilonis, A. (2011). Antibiotic resistance mechanisms of clinically important bacteria. Medicina. 47(3): 137-146.

Giobbia, M., Scotton, P. G., Carniato, A., Cruciani, M., Farnia, A., Daniotti, E., *et al.* (2003). Community-acquired *Klebsiella pneumoniae* bacteremia with meningitis and endophthalmitis in Italy. Int .J .Infect. Dis .7: 234-235.

Glupczynski, Y; Berhin, C.; Bauraing, C. and Bogaerts, P. (2007). Evaluation of a new selective chromogenic agar medium for detection of extended – spectrum β - lactamase- producing *Enterobacteriaceae*. J. Clin. Microbiol. 45(2): 501-505.

Gniadkowski, M. (2008). Evolultion of extended-spectrum β - lactamases by mutation. Clin. Microbiol. Infect. 14 (Suppl.1): 11-32.

Godfrey,H. and Evans,A.(2000).Mangmant of long -term urethral catheters: minimizing complications .Br.J.Nurs.9:74-81.

Gomez ,J.; Garcia Vazquez, E .and Ruiz Gomez, J.(2008). Clinical relevance of bacterial resistance : a historical approach (1982-2007) .Rev .Esp. Quimioter.21 :115-22.

Goossens , H .;Ferech ,M. ; Vander Stichele ,R. and Elseviers , M. (2005). Outpatient antibiotic use in Europe and association with resistance :a cross-national database study.Lancet.365 (9459):579-587.

Gregory,C.L.;Llata,E.;Stine,N. *et al.*(2010).Outbreak of carbapenem – resistant *Klebsiella pneumoniae* in Puerto Rico associated with a novel carbapenemase variant .Infec.Control.Hosp. Epidemiol. 31:476-484.

Grobner, S.; Linke, D.; Schutz, W.; Fladerer, C.; Madlung, J. Autenrieth, I. B.; Witte, W. and Pfeifer, Y. (2009). Emergence of carbapenem non- susceptible extended- spectrum β - lactamase- producing *Klebsiella pneumoniae* isolates at the university hospital of Tubingen, Germany. J. Med. Microbiol. 58-912-922.

Grohmann, E.; Muth, G. and Espinosa, M. (2003). Conjugative plasmid transfer in Gram-positive bacteria. Microbiol. and Mol. Biol. Rev. 67 (2): 277-301.

Gupta, N.; Limbago, B.M.; Patel, J.B. and Kallen, A. J. (2011). Carbapenem- resistant *Enterobacteriaceae*:epidemiology and prevention.Clin. Infect. Dis.53:60-67.

Hadi, Z. J. (2008). Detection of extended- spectrum beta-lactamases of *Escherichia coli* and *Klebsiells* spp. isolated from patients with significant bacteriuria in Najaf. M. Sc. Thesis. College of Medicine. Kufa University.

Hall, B. G. and Barlow, M. (2004). Evolution of the serine β-lactamases: past, present and future. Drug. Resist. Updat 7: 111-123.

Hall, B. G. and Barlow,M .(2005). Revised Ambler classification of *β*-lactamases. J .Antimicrob. Chemother 55: 1050-1051.

Hansen , D.S. ; Mestre, F .; Alberti, S. ; Hernandez Alles, S.; Alvarez, D.; Domenech – Senchez, A.; Gile, J.; Merino, S.; Tomas, J.M. and Benedi, V.J.(1999).*Klebsiella pneumoniae* lipopolysaccharide O typing: revision of prototype strains and O-group distribution among clinical isolates from different sources and countries .J.Clin .Microbiol.37 (1):56-62.

Hansen, D. S.; Aucken, H. M.; Abiola, T. and Podschun, R. (2004). Recommended test panel for differentiation of *Klebsiella* species on the basis of a trilateral interlaboratory evaluation of 18 biochemical tests. J. Clin. Microbiol. 42: 3665-3669.

Hashizume ,T. ;Ishino , F.; Nakagawa , J.; Tamaki, S. and Matsuhashi :, M. (1984). Studies on the mechanism of action of imipenem (N-formimidoyhhienamycin) *in vitro* :binding to the penicillin-binding proteins (PBPs) in *Escherichia coli* and *Pseudomonas aeruginosa* , and inhibition of enzyme activates due to the PBPs in *E.coli*.J.Antibiot (Tokyo) 37 :394-400.

Hawkey,P.M.and Livermore ,D.M.(2012).Carbapenem antibiotics for serious infections.BMJ.334-341.

Hawser, S. P.; Bouchillon, S. K.; Lascols, C.; Hackel, M.; Hoban, D. J.; Badal, R. E.; Woodford, N. and Livermore, D. M. (2011). Susceptibility of *Klebsiella pneumoniae* isolates from intra-abdominal infections and molecular characterization of ertapenem- resistant isolates. Antimicrobiol. Agents Chemother. 55 (8): 3917-3921.

Hemalatha, V.; Padma, M.; Sekar, U.; Vinodh, T. M. and Arunkumar, A.S. (2007). Detection of Amp C-beta lactamases production in *Escherichia coli* and *Klebsiella* by an inhibitor based method. Indian. J. Med. Res. 126: 220-223.

Heritier, C. ; Poirel, L.; Fournier, P- E.; Claverie, J-M.; Raoult, D. and Nordmann, P. (2005). Characterization of the naturally occurring oxacillinase of *Acinetobacter baumannii.* Antimicrob. Agents Chemother. 49:4174-4179.

Hernandez-Alles, S.; Conejo, M.; Pascual, A.; Tomas, J. M.; Benedi, V. J., and Martinez-Martinez, L. (2000). Relationship between outer membrane alterations and susceptibility to antimicrobial agents in isogenic strains of *Klebsiella pneumoniae*. J. Antimicrob. Chemother. 46: 273-277.

Holder, C. D. and Halkias, D. (1988). Relapsing, bacteremic *Klebsiella pneumoniae* meningitis in an AIDS patient. Am. J. Med. Sci. 295: 55-59.

Holt, J. G.; Krieg, N. R.; Sneath, H. A. Stanley, J. T. and Williams, S. T. (1994). Bergeys manual of determinative bacteriology. 9th. Ed., Baltimore; Wiliams and Wilkins, USA.

Hornsey,M.;Phee,L. and Wareham,D.W.(2011). A novel variant NDM-5 of the New Delhi metallo –β-lactamase in a multidrug –resistant *Esherchia coli* ST648 isolate recovered from a patient in the United Kingdom Antimicrob.Agents Chemother.5(11):5952-5954.

Hosoglu,S.; Gundes,S.; Kolayli,F.; Karadenizli,A.; Demirdag,K.; Gunaydin,M.; Altindis,M.; Caylan, R.and Ucmak,H.(2007). Extended-spectrum beta-lactamases in ceftazidime-resistant *Escherichia coli* and *Klebsiella pneumoniae* isolates in Turkish hospitals. Indian. J. Med. Microbiol.25:346-330.

Hsueh,P.R.;Teng,L.J.; Chen,C.Y.;Chen,W.H.;Yu,C.J.; Ho,S.W. and Luh, K.T.(2002).Pandrug resistant *Acinetobacter baumannii* causing nosocomial infections in a university hospital, Taiwan. Emerg.Infect.Dis.8:827-832.

Hujer, K. M.; Hujer, A. M.; Hulten, E. A.; Bajaksouzian, S.; Adams, J. M.;Donskey,C.J.;Ecker,D.J.;Massire,C.;Eshoo,M.W.;Sampath,R

Thomson,J.M.;Rather,P.N.;Craft,D.W.;Fishbian,J.T.;Ewell,A.J.; Jacobes,M.R.;Paterson,D.L. and Bonomo,R.A. (2006). Analysis of antibiotic resistance genes in multidrug resistant *Acinetobacter* sp. isolates from military and civilian patients treated at the walter reed army medical center. Antimicrob. Agents Chemother. 50(12): 4114-4123.

Husickova, V.; Cekanova, L. Chroma, M.; Htoutou-Sedlakova, M.; Hricova, K. and Kolar, M. (2012). Carriage of ESBL- and AmpC- positive *Enterobacteriaceae* in the gastrointestinal tract of community subjects and hospitalized patients in the Czech Republic.

Hussein, K.; Spercher, H.; Mashiach ,T.;Oren,I.;Kassis,I. and Finkelestein,R.(2009). Carbapenem resistance among *Klebsiella pneumoniae* isolates :risk factors, molecular characteristics , and susceptibility patterns . Infect. Control. Hosp.Epidemiol. 30: 666-671.

Ighinoba, A. O .and Osazuwa, F. (2012). Zero resistance to the carbapenems among extended spectrum beta- lactames producing *Klebsiella pneumoniae* in a Nigerian university hospital. Int. J. Biol. Pharmacy and Allied Sciences. 1(1): 79-83.

Ingram, P. R.; Inglis, T. J. J.; Vanzetti, T, R.; Henderson, B. A.; Harnett, G. B. and Murray, R. J. (2011). Comparison of methods for AmpC β - lactamase detection in *Enterobacteriaceae*. J. Med. Microbiol. 60: 715-721.

Iroha, I. R.; Oji, A. E.; Nwakaceze, A. E.; Ayogu, T. E.; Afiukwa, F. N.; Ejikeugwu, P.C. and Esimone, C.O. (2011). Strains of *Klebsiella pneumoniae* from intensive care unit producing CTX-M-15 extended spectrum beta- lactamases. Am. J. Microbiol. 2(2): 35-39. .

Izard, D.; Ferragut, C.; Gavini, F.; Kersters, K.; De Ley, J. and Leclerc, H. (1981). *Klebsiella terrigena*, a new species from soil and water. Int. J. Syst. Bacteriol. 31: 116-127.

Jacoby, G. A. (2009). AmpC β - lactamases. Clin. Microbial. Rev. 22(1): 161-182.

Jacoby, G. A.; Mills, D. M. and Chow, N. (2004). Role of β - lactamases and porins in resistance to ertapenem and other β - lactams in *Klebsiella pneumoniae*. Antimicrob. Agents Chemother. 48(8): 3203-3206.

Jacoby, G.A. and Munoz-Price, L.S. (2005). The New β-Lactamases. New England J. Med. 352: 380-91.

Jain, K., Radsak, K. and Mannheim, W. (1974). Differentiation of the Oxytocum group from *Klebsiella* by deoxyribonucleic acid hybridization. Int .J .Syst .Bacteriol .24:402-407.

Jemima, S. A. and Verghese,S. (2008). Multiplex PCR of *bla*CTX-M and *bla*SHV in the extended spectrum beta lactamase (ESBL) producing Gram- negative isolates. Indian. J. Med. Res. 128: 313-317.

Jeon, B-C.; Jeong , S. H.; Bae, I. K.; Know, S. B.; Lee, K.; Young , D.; Lee, J. H.; Song. J. S. and Lee , S. H. (2005). Investigation of a nosocomial outbreak of imipenem- resistant *Acinetobacter baumanni* producing the OXA-23 β - lactamase in Korea. J. Clin. Microbiol. 43: 2241-2245.

Johnson, A.P. Weinbren , M.J.;Ayling-Smith, B; Du Bois, S.K.; Amyes,S.G.B. and George R.C. (1992).Outbreak of infection in two hospitals caused by strains of *Klebsiella pneumoniae* resistant to cefotaxime and ceftazidime.J.Hosp .Infect. 20 :97-103.

Jones, C.H.; Tukman, M.; Keeney, D.; Ruzin, A. and Bradford , P. A (2009). Characterization and sequence analysis of extended spectrum beta -lactamase encoding genes from *Escherichia coli*, *Klebsiella pneumoniae* and *Proteus mirabilis* isolates collected during tigecycline phase 3 clinical trails . Antimicrob. Agents Chemother. 53: 465- 475.

Jones, R. N. (2010). Microbial etiologies of hospital – acquired bacterial pneumonia and ventilator- associated bacterial pneumonia. Clin. Infect. Dis. 51 (Suppl.1): S 81-S 87.

Kaczmarek, F. M., Dib-Hajj, F., Shang, W. and Gootz, T. D. (2006). High level carbapenem resistance in a *Klebsiella pneumoniae* clinical isolate is due to the combination of bla_{ACT-1} β-lactamase production, porin OmpK35/36 insertional inactivation, and downregulation of the phosphate transport porin PhoE. Antimicrob.Agents Chemother. 50(10): 3396–3406.

Karami, N.; Hannoun, C.; Adlerbeth, I. and Wold, A. E. (2008). Colonization dynamics of ampicillin – resistant *Escherichia coli* in the infantile colonic microbiota. J. Antimicrob. Chemother. 62: 703-708.

Karisik E.; Ellington M.J.; Pike R.; Palepou MF.; Mushtaq S.; Cheesbrough J. S.; Wilkinson P.; Livermore D.M. and Woodford N. (2007).Carbapenem resistance in an epidemic CTX-M-15 ß-lactamase-producing *Escherichia coli* strain in the United Kingdom. ESCMID. 1733-509.

Kent ,H.L. (1991) .Epidemiology of vaginitis. Am .J.Obstel . Gynecol. 165: 1168-1176.

Keynan, Y. and Rubinstein, E. (2007). The changing face of *Klebsiella pneumoniae* infections in the community. Int. J. Antimicrob. Agents. 30: 385-389.

Khorshidi, A.; Rohani, M. and Moniri, R. (2011). The prevalence and molecular characterization of extended- spectrum β - lactamases-producing *Klebsiella pneumoniae* isolates recovered from Kashan hospital university , Iran. J.J.M. 4(4): 289-294.

Kiratisin, P.; Apisarnthanarak, A.; Laesripa, C.and Saifon, P.(2008). Molecular characterization and epidemiology of extended – spectrum –beta- lactamase –producing *Escherichia coli* and *Klebsiella pneumoniae* isolates causing health care- associated infection in Thailand , where the CTX-M family is endemic .Antimicrob .Agents Chemother. 52:2818 -24.

Ko, W. C.; Paterson, D. L.; Sagnimeni, A. J.; Hansen, D. S.; Von Gottberg, A.; Mohapatra, S.; Casellas, J. M.;Goossens ,H.;Mulazimoglu,L;Trenholme,G.;Klugman,K.P.;McCormack,J. G. and.Yu, V.L. (2002). Community - acquired *Klebsiella pneumoniea* bacteremia: global differences in clinical patterns. Emerg. Infect. Dis .8: 160-166.

Koch,A.L.(2000).Penicillin bindings proteins,beta-lactams, and lactamases:offensives attacks and defensive countermeastures . Crit.Rev.Microbial.26(4):205-220.

Kochar,S;Sheard,T. and Sharma,R.(2009).Success of an infection control program to reduce the spread of carbapenem –resistance *Klebsiella pneumoniae*.Infect. Control Hosp.Epidemiol.30:447-452.

Koh,T. H. ; Li-Hweisng ; Babini G, S. ; Woodford , N. ; Livermore , D .and Hall , L.M.C. (2001). Carbapenem –resistant *Klebsiella pneumoniae* in Singapore producing IMP-1 beta-lactamase and lacking on outer membrane protien.Antimicrob.Agent Chemother.45(6) :1939-1940.

Kollef, M. H. and Fraser, V. J. (2001). Antibiotic resistance in intensive care unit. Ann .Intern .Med. 134(40):298-314.

Koneman, E.W.; Allen, S.D. ; Janda, W. M.; Schreckenberger, P.C. and Winn, W.C. The *Enterobacteriaceae*. In: Koneman, E.W.; Allen , S.D. Janda, W.M.; Schreckenberger, P.C. and Winn, W.C. (1994). Introduction to diagnostic microbiology .Philadelphia :J.B Lippincott.

Kosmidis, C.; Poulakou, G.; Markogiannakis, A. and Daikos, G. L. (2012). Teratment options for infections caused by carbapenem-resistant Gram-negative bacteria. Eur. Infect. Dis. 6(1): 28-34.

Ktari, S.; Artlet, G.; Mnif, B.; Gautier, V.; Mahjoubi, F.; Jmeaa, M. B.; Bouaziz, M. and Hammami, A. (2006). Emergence of multidrug-resistant *Klebsiella pneumoniae* isolates producing VIM-4 metallo- β - lactamase, CTX- M-15 extended – spectrum-β - lactamase, and CMY-4 AmpC β - lactamase in a Tunisian university hospital . Antimicrob. Agents Chemother .50 (12): 4198-4201.

Kucukates, E. and Kocazeybek, B.(2002). High resistance rate against 15 different antibiotics in aerobic Gram- negative bacterial isolates of cardiology intensive care unit patients .Ind .J. Med. Microbiol. 20 (4) :208 -210.

Kumar, M.; Behera, B.; Sajiri, S.S.; Pal, K.; Kay, S.S. and Roy, S. (2011). Bacterial vaginosis: etiology and modalities of treatment – abrief note. J. Pharm- Bioallied. 3(4): 489. 503.

Kumarsamy, K.K.; Toleman, M. A.and Walsh, T. R. (2010). Emergence of a new antibiotic resistance mechanism in India , Pakistan, and the UK: a molecular, biological, and epidemiological study . Lancet. Infect. Dis. 10: 597-602.

Kuo,L.C.;Yu,C.J.;Lee,L.N.;Wang,J.L.;Wang,H.C.;Hseuh,P.R and Yang, P.C .(2003). Clinical features of pandrug resistant *Acinetobacter baumannii* bacteremia at a university hospital in Taiwan.J.Formos.Med.Assoc.102:601-606.

Kwak ,Y.G.; Choi, S.H.; Choo,E.J *et al* .(2005).Risk factors for the acquisition of carbapenem resistant *Klebsiella pneumoniae* among hospitalized patients. Microb. Drug. Resist .11 :165 -9.

Lagace- Wiens, P. R. F.; Tailor, P.; Simmer, M.; Decorby, J. A.; Karlowsky, A.; Walkty, D. J.; Hoban and Zhand, G. G. (2011). Activity of NXL104 in combination with beta- lactams against genetically characterized *Escherichia coli* and *Klebsiella pneumoniae* isolates producing class A extended- spectrum beta-lactamases and class C beta- lactamases. Antimicrob. Agents Chemother. 55: 2434-2437.

Lederman, E. R.and Crum , N. F. (2005). Pyogenic liver abscess with a focus on *Klebsiella pneumoniae* as primary pathogen: an emerging disease with unique clinical characteristics. Am. J. Gastroenterol. 100: 322-331.

Lee, K.; Chong, Y.; Shin, H. B.; Kim, Y. A.; Yong, D. and Yum, J. H. (2001). Modified Hodge and EDTA- disc Synergy tests to screen metallo- β - lactamase- producing strains of *Pseudomonas* and *Acinetobacter* species. J. Clin. Microbiol. Infect. 7: 88-91.

Lee, K.; Lim, Y. S.; Yong , Yum, J. H. and Chong, Y. (2003).Evaluation of the Hodge test and the imipenem-EDTA double-disk synergy test for differentiating metallo- β - lactamase- producing isolates of *Pseudomonas* spp. and *Acinetobacter* spp. J. Clin. Microbiol., 41(10): 4623-4629.

Lee, K.; Yum, J. H.; Yong, D.; Lee, H. M.; Kim, H. D.; Docquier, J. D.;Rossolini,G.M. and Chong,Y. (2005). Novel acquired metallo-β -lactamase gene, *bla*(SIM-1), in a class 1 integron from *Acinetobacter baumannii* clinical isolates from Korea. Antimicrob. Agents Chemother. 49: 4485-4491.

Levinson, W. and Jawetz, E. (2000). Medical microbiology and Immunology. 4th edition. Appleton and Lange

Li,B;Yi.Y. ; Wang,Q. ; Woo,P.C.Y .; Tan,L. ; Jing,H.; Gao,G.F. and Liu,C.H. (2012). Analysis of drug resistance determinants in *Klebsiella pneumoniae* isolates from a tertiary – care hospital in Beijing,China.Plos one.7:e42280.

Li, C. K.; Shing, M. M.; Chik, K. W.; Lee, V. and Yuen, P. M. (2001). *K. pneumoniae* meningitis in thalassemia major patients. Pediatr. Hematol .Oncol. 18: 229-232.

Lim, K. T.; Yeo, C. C.; Yasin, R. M. Balan , G. and Thong, K. L. (2009). Characterization of multidrug-resistant and extended- spectrum β - lactamase – producing *Klebsiella pneumoniae* strains from Malaysian hospitals. J. Med. Microbiol. 58:1463-1469.

Limbago, B. M.; Rasheed, J. K.; Anderson, K. F.; Zhu, W.; Kitchel, B.; Watz, N.; Munro, S.; Gans, H.; Banaei, N. and Kallen, A. J. (2011). IMP-producing carbapenem resistant *Klebsiella pneumoniae* in the United States. J. Clin. Microbiol. 49(12): 4239-4245.

Livermore, D. M. (1995).Beta-lactamases in laboratory and clinical resistance. Clin.Microbiol.Rev.8:557-584.

Livermore D.M.; Canton,R.; Gniadkowski,M.; Nordmann,P.; Rossolini, .M.;Arlet,G.;Ayala,J.;Coque,T.M.;Kernzdanwicz,I.;Luzzaro,F.; Poriel,L.and Woodford,N. (2007).CTX-M:changing the face of ESBLs in Europe.J.Antimicrob.Chemother.59.156-174.

Livermore, D.M.;Hope ,R.; Fagan, E.J; Warner, M.; woodford, N. and Potz, N. (2006).Activity of temocillin against prevalent ESBL- and AmpC- producing *Enterobacteriaceae* from south-east England.J.Antimicrob. Chemother.57:1012-1014.

Livermore D.M. and Brown, D.F.J. (2001) .Detection of β- Lactamase-mediated resistance .J. Antimicrob. Chemother., 48 (Supl 1) : 59-64.

Livermore, D. M. and Williams, J. D. (1996). Mode of action and mechanisms of bacterial resistance. In V. Lorian (Ed.), Antibiotics in Laboratory Medicine, 4th Ed. New York, Williams and Wilkins, Baltimore 502-578.

Livermore, D. M. and Woodford, N. (2006). The β - lactamase threat in *Enterobacteriaceae*, *Pseudomonas* and *Acinetobacter*. Trend. Microbiol. 14(9): 413-420.

Livermore, D.M.; Mushtaq, S, and Warner, M.(2010).Activity of BAL30376 (monobactum BAL1976+BAL29880+clavulanate) versus Gram-negative bacteria with characterized resistance mechanisms J. Antimicrob. Chemother.65:2382-2395.

Livermore, D. M.; Mushtaq, S.; Warner, M.; Zhang, J.; Maharjan, S.; Doumith, M. and Woodford, N. (2011). Activities of NXL104 combinations with ceftazidime and aztreonam against carbapenemase- producing *Enterobacteriaceae*. Antimicrob. Agents Chemother. 55(1): 390-394.

Loli,A. ; Tzouvelekis,L.S ; Tzelepi,E.;Carattoli,A. ; Vatopoulos,A.c.; Tassios,P.T. and Miriagou,V.(2006).Sources of diversity of carbapenem resistance levels in *Klebsiella pneumoniae* carrying VIM-1.J.Antimicrob.Chemother.58:669-672.

Luzzaro, F.; Docquier, J. D.; Colinon, C.; Endimiani, A.; Lombardi, G.; Amicosante, G.; Rossolini, G. M. and Toniolo, A. (2004). Emergence in *Klebsiella pneumoniae* and *Enterobacter cloacae* clinical isolates of VIM -4 metallo - β - lactamase encoded by a conjugative plasmid. Antimicrob. Agents Chemother. 48(2). 648-650.

Lytsy, B.; Sandegren, L.; Tano, E.; Torell, E.; Andersson, D. I. and Melhus, A. (2008). The first major extended-spectrum β-lactamase outbreak in Scandinavia was caused by clonal spread of a multiresistant *Klebsiella pneumoniae* producing CTX-M-15. APMIS. 116: 302-308.

MacFaddin, J. F. (2000). Biochemical tests for identification of medical bacteria. 3rd ed. Lippincott Williams and Wilkins, USA.

Magiorakos, A.P.; Srinivasan, A.; Carey, R. B.; Carmeli, Y.; Falagas, M. E.; Giske, C. G.; Harbarth, S.; Hindler, J. F.; Kahlmeter, G.; Olsson-Liljequist, B.; Paterson, D. L.; Rice, L. B.; Stelling, J.; Struelens, M. J.; Vatopoulos, A.; Weber, J. T. and Monnet, D.L. (2011). Multidrug-resistant ,extensively drug-resistant and pandrug-resistant bacteria: an international expert proposal for interim standard definitions for acquired resistance Clin. Microbiol. Infect.18:268-281.

Manchanda, V. and Singh, N.P. (2003). Occurrence and detection of AmpC β - lactamases among Gram- negative clinical isolates using a modified three- dimentional test at Guru Tegh Bahadur Hospital, Delhi. India. J. Antimicrob. 51:415-418.

Manoharan, A.; Sugumar, M.; Kumar, A.; Jose, H.; Mathai, D. and ICMR-ESBL study group (2012). Phenotypic and molecular characterization of AmpC β - lactamases among *Escherichia coli*, *Klebsiella* spp. and *Enterobacter* spp. from five Indian Medical Centers. Indian. J. Med. Res. 135: 359-364.

Marsik, F. J. and Nambiar, S. (2011). Review of carbapenemases and AmpC- beta lactamases. Pediatr. Infect. Dis. J. 30 (12): 1094-1095.

Martinez,P.; Garzon,D. and Matter,S. (2012). CTX-M producing *Escherichia coli* and *Klebsiella pneumoniae* isolates from community –acquired urinary tract infections in Valledupar Colombia.PRAZ.Infect.Dis.16(5).420-425.

Mathur, P.; Tatman, A.; Das, B.and Dhawan, B. (2002). Prevalence of extended beta lactamase producing Gram negative bacteria in a tertiary care hospital. Indian .J .Med. Res. 115 :153-7.

Matsumoto, Y.; Ikeda, F.; Kamimura, T.; Yokota, Y. and Mine, Y. (1988). Novel plasmid-mediated β-lactamase from *Escherichia coli* that inactivates oxyimino-cephalosporins. Antimicrob .Agents Chemother. 32: 1243-1246.

McCallum, K. L. ; Schoenhals,G.; Laakso,D.; Clarke ,B. and Whitfield,C. (1989) . A high molecular weight fraction of smooth

Lipopolysaccharide in *Klebsiella* serotype O1 : K20 contains a unique O antigenic epitope and determines resistance to non-specific serum killing. Infect. Immum. 57:3816 -3822.

McDermott, P.F.; Walker, R.D.and White, D.G. (2003). Antimicrobials: modes of action and mechanisms of resistance. Inter. .J . Toxi. 22: 135-143.

Medeiros, A.A. (1997).Evolution and dissemination of β -lactamases accelerated by generations of β -lactam antibiotics.Clin.Infect . Dis. 24: S 19-S 45.

Michalopoulos, A.S.; Tsiodras, S. ; Rellos,K.; Mentzelopoulos ,S. and Falagas ,M.E. (2005) . Colistin treatment in patients with ICU – aquired infections caused by multiresistant Gram- negative bacteria: The rehaissance of an old antibiotic.Clin .Microbiol. Infect .11 :115 -121.

Miranda,G.;Castro,N. ;Leanos ,B.;Valenzuela ,A.; Garze –Ramos ,U.;Rojas,T.;Solorzono,F.; Chihu,L and Silva,J.(2004).Clonal and horizontal dissemination of *Klebsiella pneumoniae* expressing SHV-5 extended-spectrum β -lactamase in a Mexican pediatric hospital.J.Clin .Microbiol.42(1):30 -35.

Miriagou, V.; Carattoli, A.; Tzelepi, E.; Villa, L. and Tzouvelekis, L. S. (2005).IS*26*-associated *In4*-type integrons forming multiresistance loci in enterobacterial plasmids. Antimicrob. Agents Chemother .49: 3541-3543.

Misra, R. N. (2012). Metallo β - lactamases: a prespective and implications. Med. J. 5: 10-13.

Mohamudha, P. R.; Harish, B. N. and Parija, S. C. (2012). Molecular description of plasmid- mediated AmpC β - lactamases among nosocomial isolates of *Escherichia coli* and *Klebsiella pneumoniae* from six different hospitals in India. Indian. J. Med. Res. 135: 114-119.

Mohamudha, P. R.; Srinivas, A. N.; Rahul, D.; Harish, B. N. and Parija, S. C. (2010). Molecular epidemiology of multidrug resistant extended- spectrum β - lactamase producing *Klebsiella pneumoniae* outbreak in a neonatal intensive care unit. Inter. Collabor. Res. Inter. Med. Public Health. 2(7): 226- 237.

Moland, E. S.; Hanson,N.D, ; Herrera,V.K.; Black,J.A.; Lockhart, T.J.; Hossain,A.; Johnson,A.J.; Goering,R.V. and Thomson,K.S. (2003). Plasmid-mediated, carbapenem-hydrolysing β - lactamase, KPC-2, in *Klebsiella pneumoniae* isolates. J. Antimicrob . Chemother. 51:711-714.

Moosavian, M. and Deiham, B. (2012). Distribution of TEM, SHV and CTX-M genes among ESBL- producing *Enterobacteriaceae* isolates in Iran. Afr. J. Microbiol. Res. 6(26): 5433-5439.

Mullany S. (2002). Cellular and molecular life sciences .Birkhäuser Basel, Spain.

Mulvey, M. R.; Bryce, E. Boyd, D. A.; Ofner- Agostini, M.; Simor, A. E. and Paton, S. (2005). Molecular characterization of cefoxitin-resistant *Escherichia coli* from Canadian hospitals. J. Antimicrob. Agents Chemother. 49 (1): 358-365.

Munoz- Price, L. and Quinn, J. P. (2009). The spread of *Klebsiella pneumoniae* carbapenemases: a tale of strains, plasmids., transpons. Clin. Infec. Dis. 49: 1739-1741.

Mushtaq, S.; Warner, M.; Williams, G.; Critchley, I. and Livermore, D. M. (2010). Activity of chequerboard combinations of ceftaroline and NXL104 versus β - lactamase- producing *Enterobacteriaceae* J. Antimicrob. Chemother. 65: 1428-1432.

Mumtaz, S.; Ahmad , M.; Aftab, I.; Akhtar, N.; Hassan, M. and Hamid, A. (2008). Aerobic vaginal pathogens and their sensitivity pattern. J. Ayub. Med. Coll. Abbottabad. 20(1): 113-117.

Mylonas, I. and Friese ,K. (2007). Genital discharge in women. MMW Frotschr. Med.149(35-36):42-46.

Najmadeen, H.H. (2006). Prevalence of *Klebsiella* spp. as a nosocomial bacteria in Sulaimani hospitals. M.SC. Thesis. College of Sciences .University of Salahaddin.

National Committee for Clinical Laboratory Standards (NCCLS). (2003b). Performance standards for antimicrobial disc susceptibility testing. Disc diffusion. 8th ed. Informational supplement. M100-S13. NCCLS. Wayne, Pa.

Neuner, E. A.; Yeh, J. Y.; Hall, G. S.; Sekeres; Endimiani, A.; Bonomo, R. A.; Shrestha , N. K.; Fraser, T. G. and Duin, D. V. (2011). Treatments and outcomes in carbapenem- resistant *Klebsiella pneumoniae* bloodstream infections. Diagn. Microbiol. Infect. Dis. 69(4): 357-362.

Nguyen Thi, P. L.; Yassibanda, S.; Aidara, A.; Le Bouguenec, C. and Germani, Y. (2003). Enteropathogenic *Klebsiella pneumoniae* HIV-infected adults, Africa. Emerg .Infect. Dis .9: 135-137.

Nian, H.; Chu, Y.; Tian, S.; Hua, D.; Ding, L.; Guo, L. and Shang, H. (2012). Study on capsular serotype of *Klebsiella pneumoniae* from patients with liver abscesses and its clinical characteristics. Afr. J. Microbiol. Res. 6(14): 3523-3527.

Nordmann, P.; Cuzon, G. and Naas, T. (2009). The real threat of *Klebsiella pneumoniae* carbapenemase- producing bacteria. Lancet. Infect. Dis. 9: 228-236.

Nordmman, P.;Poril,L.;Carrer,A.:Toleman,M.A.and Walsh ,T.R. (2011). How to detect the NDM-1 producers .J.Clin .Microbiol .44(9):3139-3144.

Normark, H.B. and Normark, S. (2002). Evolution and spread of antibiotic resistance. J .Intern .Med. 252: 91-106.

Obiajuru, I.O.C; Chukuezi, A .B .and Okechi, O.O. (2010). Prevalence of AmpC and extended spedrum β- lactamse producing bacteria in clinical bacterial isolates at Imo State University Wospital , Orlu , Nigeria. Nigerian J. Microbial .24(1) : 2062- 2067.

Ohmori,S.; Shiraki,K.; Ito,K.;Inoue,H.; Ito,T.;Sakai,T.; Takase,K.and Nakano,T.(2002). Septic endophthalmitis and meningitis associated with *Klebsiella pneumoniae* liver abscess. Hepatol.Res.22:307-312.

Otman, J.;Perugini, M.E.;Cavasin, E. and Vidotto, M.C. (2002).Study of an outbreak of extended –spectrum beta lactamase-producing

Klebsiella species in a neonatal intensive care unit in Brazil by means of repetitive intergenic consensus sequence – based PCR (ERIC-PCR).Infect.Contr.Hosp.Epidem.23:8-9.

Overturf, G.D. (2010). Carbapenemases: a brief review for pediatric infectious disease specialists. Pediatri. Infect. Dis. J. 29: 68-70.

Ozyilmaz, E.; Akan, O.A.; Gulhan, M.; Ahmed, K.and Nagatake, T. (2005). Major bacteria of community-acquired respiratory tract infections in Turkey. Jpn. J .Infect. Dis. 58: 50-52.

Pai, H.; Kang,C.I.;Byeon,J.H.;K.D.;Park,W.B and Kim,H.B. (2004). Epidemiology and clinical features of bloodstream infections caused by AmpC- type-beta –lactamase producing *Klebsiella pneumoniae.*Antimicrob.Agents Chemother.48:3720-3728.

Pallasch, T. J. (2003). Antibiotic resistance. Dent. Clin. N. Am. 47: 623-639.

Panagea,T. ;Galani,I.; Souli,M. ;Adamou,P.A; Antoniadiou,A. and Giamarellou,H.(2011).Evalution of carbapenemase –producing *Enterobacteriaceae* in rectal surverillance cultures .Int. J.Antimicrob.Agent 37:124-148.

Pangon, B.; Bizet, C.; Bure, A.; Pichon, F.; Philippon, A.; Regnier, B. and Gutmann, L. (1989). In vivo of a cephamycin resistant, porin-deficient mutants of *Klebsiella pneumoniae* producing a TEM-3 β - lactamase. J. Infect- Dis. 159:1005-1006.

Papp-WallaceK.M;Endimiani,A;Taracila,M.A.and Bonomo, R.A.(2011). Carbapenem past,present and feature.Antimicrob.Agents Chemother.55(11):4943-4960.

Park, Y.; Kang, H- K.; Bae, K.; Kim, J.; Kim, J-S.; Uh, Y.; Jeong, S. H. and Lee, K. (2009). Prevalence of the extended- spectrum β - lactamase and *qnr* genes in clinical isolates of *Escherichia coli* . Korean. J. Lab. Med. 29: 218-223.

Parveen, M.; Harish, B. N. and Parija, S. C. (2010b). AmpC beta – lactamases among Gram negative clinical isolates from a tertiary hospital South India. Bra. J. Microbiol. 41: 596-602.

Parveen, R. M.; Harish, B. N. and Parija, S. C. (2010a). Emerging carbapenem resistance among nosocomial isolates of *Klebsiella pnenmoniae* in South India. Inter. J. Pharma. Bio. Sci. 1(2):1-12.

Pasteran, F.; Mendez, T.; Guerriero, L.Rapoport,M and Corso,A. (2009). Sensitive screening tests for suspected class A carbapenemase production in species of *Enterobacteriaceae*. J. Clin .Microbiol: 47:1631-1639.

Patel, N.; Harrington, S.; Dihmess, A.; Woo, B.; Masoud, R.; Martis, P. ; Fiorenza, M.; Graffunder, E.; Evans, A.; MeNutt, L-A and Lodise,T. P. (2011). Clinical epidemiology of carbapenem-intermediate or resistant *Enterobacteriaceae*. J. Antimicrob. Chemother . 66: 1600-1608.

Patel, T.K.; Sheth, K. V. and Tripathi, C. B. (2012). Antibiotic sensitivity pattern im neonatal intensive care unit of a tertiary care hospital of India. Asian. J. pharm. Clin. Res. 5: 46-50.

Patel, J. B.; Rasheed,J.K. and Kitchel,B. (2009). Carbapenemases in *Enterobacteriaceae*: activity, epidemiology, and laboratory detection. Clin. Microbiol. Newsl. 31:55-62.

Patel, G., and Bonomo,R.A. (2011). Status report on carbapenemases: challenges and prospects. Expert Rev. Anti. Infect. Ther. 9:555-570.

Patel,G.; Huprikar,S.; Factor,S.H.; Jenkins,S.G.and Calfee,D.P. (2008) .Outcomes of carbapenem- resistant *Klebsiella pneumoniae* infection and the impact of antimicrobial and adjunctive therapies. Infect. Control. Hosp. Epidemiol .29: 1099-1106.

Paterson, D.L. (2006). Resistance in Gram-Negative Bacteria: *Enterobacteriaceae*. American J .Med. 119: S20-S28.

Paterson, D. L. and Bonomo, R. A. (2005). Extended-spectrum β-lactamases: a clinical update. Clin. Microbiol. Rev .18: 657-686.

Paterson, L. D.; Hujer, K. M.; Hujer, A. M.; Yeiser, B. and Michael, D. (2003). Extended- spectrum β - lactamases in *Klebsiella pneumoniae* bloodstream isolates from seven countries: dominance and widespread prevalence of SHV and CTX-M type β - lactamases. J. Antimicrob. Agents Chemother. 47: 3554-3560.

Pathak, A.; Marothi, Y. ; Kekre, V.; Mahadik, K.; Macaden, R. and Lundborg, C. S. (2012). High prevalence of extended- spectrum β - lactamase-producing pathogens: results of a surveillance study in two hospitals in Ujjain, India. Infect. Drug. Resist. 5: 65-73.

Paton, R.; Miles, R. S.;Hood, J.; Amyes, S. G.; Miles, R. S. and Amyes, S. G. (1993). ARI 1: β-lactamase-mediated imipenem resistance in *Acinetobacter baumannii*. Int .J. Antimicrob .Agents. 2: 81-87.

Paulson,O.S.(2008). Biostatistics and Microbiology:ASurvival Manual.Springer Sience and BusinessMedia,LLC.

Payne, D. J. ; Cramp , R .; Winstanley, D.J. and Knowles , D.J. (1994) . Comparative activities of clavulanic acid , sulbactam , and tazobctam against clinically important β- lactamases. Antimicrob. Agents Chemother .38 :767-772.

Peirano, G.; Seki, L. M. ; Passos, V. L.V; Pinto, M. C. F. G.; Guerra, L. R. and Asensi, M. D. (2009).Carbapenem –hydrolyzing β - lactamase KPC-2 in *Klebsiella pneumoniae* isolated in Rio de Janeiro,Brazil. J. Antimicrob. Chemother. 63: 265-268.

Perez, F.; Endimiani, A.; Ray, A. J.; Decker, B. K.; Wallace, C. J.; Hujer, K. M.; Ecker, D. J.; Adams, M. D.; Toltzis, P.; Dul, M. J.; Windau, A.; Bajaksouzian, S.; Jacobs, M. R.; Salata, R. A. and Bonomo, R. A. (2010). Carbapenem-resistant *Acinetobacter baumannii* and *Klebsiella pneumoniae* across a hospital system: impact of post- acute care facilities on dissemination. J. Antimicrob. Chemother.1-12.

Pfeifer, Y. ; Cullik ,A. and Witte ,W. (2010) . Resistance to cephelosporins and carbapenems in Gram- negative bacterial pathogens. Int.J. Med .Microbiol 300 :371 -379.

Pfeifer,Y.; Wilharm ,G.;Zander,E.,Wichelhaus,T.A.;Gottig,S.;Hunfeld,K. P.; Seifer,H.; Witte,W. and Higgins, G.P.(2011).Molecular characterization of bla_{NDM-1} in *Acintobacter baumannii* strain isolated in Germany in 2007.J.Antimicrob.Chemother.66(9).1998

-2001.

Philippe ,E. ;Weiss, M. ;Shultz, JM.; Yeomans, F. and Ehrenkranz, NJ.(1999). Emergence of highly antibiotic- resistant *Pseudomonas aeruginosa* in relation to duration of empirical antipseudomonal antibiotic treatment. Clin. Perform. Qual. Health Care. 7: 83-7.

Philippon, A.; Arlet, G. and Jacoby, G. A. (2002). Plasmid- determined AmpC-type β - lactamases. J. Antimicrob. Agents Chemother. 46:1-11.

Pitout,J.D. and Laupland ,K.B.(2008).Extended spectrum-β-lactamase-producing *Enterobacteriaceae:* an emergingpublic-health problem concern.Lancet.Infect.Dis.8:159-166.

Podschun, R. and Sahly , H.(1991).Hemagglutinins of *Klebsiella pneumoniae* and *Klebsiella oxytoca* isolated from different sources.Zbl.Hyg . Umweltmed.191(1):46-52.

Podschun, R. and Ullmann, U. (1998). *Klebsiella* spp. as nosocomial pathogens: epidemiology, taxonomy, typing methods, and pathogenicity factors. Clin. Microbiol. Rev. 11: 589-603.

Poirel, L.; Thomas,L.I.; Naas,T.; Karim,A. and Nordmann,P. (2000). Biochemical sequence analyses of GES-1, a novel class A extended- spectrum β - lactamase, and the class I integron In52 from *Klebsiella pneumoniae*. Antimicrob. Agents Chemother. 44: 622-632.

Poirel, L. ; Abdelaziz,M.O. ; Bernabeu,S. and Nordmann,P. (2013). Occurrence of OXA-48 and VIM - carbapenemase-producing *Enterobacteriaceae* in Egypt.Inter.J.Antimicrob.Agents .41:90-95.

Poirel, L.; Al- Maskari, Z.; Al- Rashdi, F.; Bernaben , S. and Nordmann, P. (2011). NDM-1 producing *Klebsiella pneumoniae* isolated in the Sultanate of Oman. J. Antimicrob. Chemother. 66:304-306.

Poirel, L.; Weldhagen, G.F.; Naas, T.; De Champs, C.; Dove, M.G.; Nordmann, P. (2001). GES-2, a class A beta-lactamase from *Pseudomonas aeruginosa* with increased hydrolysis of imipenem. Antimicrob. Agents Chemother. 45 (9): 2598-2603.

Poirel, L.; Weldhagen, G.F.; De Champs, C. and Nordmann, P.A. (2002). Nosocomial outbreak of *Pseudomonas aeruginosa* isolates expressing the extended-spectrum beta-lactamase GES-2 in South Africa. J. Antimicrob. Chemother. 49 : 561-5.

Polsfuss, S.; Bloemberg, G. V.; Giger, J.; Meyer, V.; Bottger, E. C. and Hombach, M. (2011). Practical approach for reliable detection of AmpC beta- lactamase- producing *Enterobacteriaceae*. J. Clin. Microbiol. 49(8): 2798-2803.

Pospiech, T. and Neumann, J. (1995). In genomic DNA isolation Kieser eds. John Innes Center. Norwich NR4 7UH. U.K.

Prakask, S. (2006). Carbapenem sensitivity profile amongst bacterial isolates from clinical specimens in Kanpur city. Indian J. Crit. Care. Med. 10: 250-253.

Queenan, A. M. and Bush, k. (2007). Carbapenemases: the versatile β - lactamases. J. Clin. Microbiol. Rev. 20 (3): 440-458.

Queenan, A. M.; Torres- Viera, C.; Gold, H. S.; Carmeli, Y.; Eliopoulos, G. M.; Moellering, R. C.; Quinn, J. P.; Hindler, J.; Medeiros, A. A. and Bush, K. (2000). SME-type carbapenem- hydrolyzing class A β - lactamases from geographically diverse *Serratia marcescens* strain. Antimicrob. Agents Chemother. 44(11): 3035-3039.

Radji, M.; Fauziah, S. and Aribinuko, N. (2011). Antibiotic sensitivity pattern of bacterial pathogens in the intensive care unit of Fatmawati Hospital, Indonesia. Asian. Paci. J. Trop. Biomed. : 39-42.

Rampure , R.; Gangane , R.; Oli, A.K . and Chandrakanths , R. (2013).Prevalence of MDR- ESBL producing *Klebsiella pneumoniae* isolated from clinical samples .J. Microbiol. Biotech. Res. 3 (1) :32 -39.

Rennie, R. P.; Anderson, C. M.; Wensley, B. G.; Albritton, W. L. and Mahony, D. E. (1990). *Klebsiella. pneumoniae* gastroenteritis masked by *Clostridium perfringens*. J. Clin. Microbiol. 28: 216-219.

Rice ,L.B.(2000).Bacterial monopolists: the bundling and dissemination of resistance genes in Gram-positive bacteria. Clin. Infect. Dis.31(3):762-769.

Rice, L. B.; Yao, J. D.; Klimm, K.; Eliopoulos, G. M. and Moellering, R. C. (1991). Efficacy of different β - lactams against an extended-spectrum β - lactamase- producing *Klebsiella pneumoniae* strain in the rat intra- abdominal abscess model. J. Antimicrob. Agents Chemother. 35: 1243-1244.

Richards, M. J.; Edwards , J. R.; Cluver, D. H.; Gaynes, R.P. and the National Nosocomial Infections Surveillance System. (2000) Nosocomial infections in combined medical- surgical intensive care units in the United States. Infect. Control Hosp. Epidemiol. 21: 510-515.

Robin, F.; Hennequin, C.; Gniadkowski, M.; Beyrouthy, R.; Empel, J.; Gibold, L. and Bonnet, R. (2011). Virulence factors and TEM-Type β - lactamases produced by two isolates of an epidemic *Klebsiella pneumoniae* strain. Antimicrob. Agents Chemother. 56(2): 1101-1104.

Robledo,I.E.; Vasquez,G.J.: Moland,E.S.; Aquino,E.E.;Goering,R.V.; Thomoson ,K.S.;Sante,M.I.and Hanson, N.D .(2011). Dissemination and molecular epidemiology of KPC-producing *Klebsiella pneumoniae* collected in Puerto Rico medical center hospitals during a1- year period. Epidemiol. Res.Inter. 1-8.

oh,K.H.:Lee,C.K.;Sohn,J.W.;Song,W.;Yong,D.and Lee,K. (2011). Isolation of *Klebsiella pneumoniae* isolate of sequence type 258 producing KPC-2 carbapenemase in Korea. Korean .J.Lab.Med.31:298-301.

Roy,S.;Viswanathan,R.;Singh,A.K.;Das,P.and Basulla ,S.(2011).Sepsis in neonaets due to imipenem –resistant *Klebsiella pneumoniae* producing NDM-1 in India.J.Antimicrob.Chemother.66 (6):1411-1413.

Rubio- Perez, I.; Martin- Perez, E.; Carcia, D. D.; Calvo, M. L- B and Barrera, E. L. (2012). Extended- spectrum beta- lactamase-producing bacteria in a tertiary care hospital in Madrid:

epidemiology, risk factors and antimicrobial susceptibility patterns. Emerg. Health. Threats. J. 5:11589-11594.

Rudresh, S.M. and Nagarathnamma, T. (2011). Two simple modifications of modified three- dimensional extract test for detection of AmpC β - lactamases among the members of family *Enterobacteriaceae*. Chron. Young Sci. 2: 42-46.

Sacha, P.; Ostas, A.; Jaworowska, J.; Wieczorek, P.; Ojdana, D.; Ratajczak, J.and Tryiszewska, E. (2009). The KPC type β - lactamases: new enzymes that confer resistance to carbapenems in Gram- negative bacilli. Folia. Histochem. Cytobiol. 47(4): 537-543.

Sacha,P.T.;Ojdana,D.; Wieczorek ,P.;Klowska, W.;Krawczk,M.; Czaban,S.; Oldak,E .and Tryniszewska,E.(2012).Genetic similarity and antimicrobial susceptibility of *Klebsiella pneumoniae* –producing carbapenemase (KPC-2) isolated in different clinical specimens received from university hospitals in Notherastern Poland.Afr.J.Microbiol.6(41):6888-6892.

Sakazaki, R.; Tamura, K.; Kosako, Y. and Yoshizaki, E. (1989). *Klebsiella ornithinolytica* sp. nov., formerly known as ornithinepositive *Klebsiella oxytoca*. Curr. Microbiol. 18: 201-206.

Sakharkar,M.K.; Jayaraman,P. ;Soe,W.M.; Chow,V.T.; Sing,L.C.and Sakharkar,K.R.(2009).In vitro combinations of antibiotics and phytochemicals against *Pseudomonas aeruginosa* .J.Microbiol. Immunol.Infect.42(5):364-370.

Samatha, P. and Praveen , K. V. (2011). Prevalence of ESBL and AmpC β - lactamases in Gram- negative clinical isolates. J. Biosci. Tech 2(4): 353-357.

Sambrook, J. ; Fritshch, E.F. ; and Maniatis, T. (1989). Molecular cloning: a laboratory manual, 2nd ed. Cold spring Harbor Laboratory Press, Cold Spring Harbor, N. Y.

Sambrook, J., and Russell, D.W. (2001). Molecular cloning: laboratory manual, 3rd ed. Cold Spring Harbor Laboratory Press, Cold Spring Harbor, N.Y.

Samuelsen,Q. ; Toleman , M.A.; Hasseltedt, V.; Fnursted , K ; leegaard , T.M ;Walsh,T,R. Sundsfjord, A , and Giske , G.C. (2011). Molecular characterization of VIM -producing *Klebsiella pneumoniae* from Scandinavia reveals genetic relatedness with international clonal complexes encoding transferable multidrug resistance . Clin .Microbiol .Infect .17 : 1811- 1816.

Sanchez-Romero, I.; Asensio, A.; Oteo, J.; Munoz-Algarra, M.; Isidoro, B.; Vindel, A.; Alvarez- Avello, J.; Balandin- Moreno, B.; Cuevas, O.; Fernandeez- Romero, S.; Azanedo, L.; Saez, D. and Campos, J. (2011). Nosocomial outbreak of VIM-1- producing *Klebsiella pneumoniae* isolates of multilocus sequence type 15: molecular basis, clinical risk factors, and outcome. Antimicrob. Agents Chemother. 56 (1): 420-427.

Sarojamma, V. and Ramakrishna, V. (2011). Prevalence of ESBL-producing *Klebsiella pneumoniae* isolates in tertiary care hospital. I.S.R.M. Microbiol.

Sauvage,E.;Kerff,F. *et al.* (2008).The pencillin binding proteins: structure and role in peptidoglycan biosynthesis. FEMS Microbiol. Rev.32:234-258.

Schneider, I.; Queenan, A. M. and Bauernfeind , A. (2006). Novel carbapenem- hydrolyzing oxacillinase OXA-62 from *Pandoraea pnomenusa*. Antimicrob. Agents Chemother. 50: 1330-1335.

Schwaber, M.J. and Carmeli, Y.(2008). Carbapenem-resistant *Enterobacteriaceae*: a potential threat. JAMA. .300(24):2911.

Schwaber,M.J.; Klarfeld-Lidji, S.; Navon- Venezia,S. *et al.* (2008).Predictors of carbapenem-resistant *Klebsiella pneumoniae* acquisition on mortality. Antimicrob. Agents Chemother. 52:1028-1033.

Schmidtke,A,J.and Hanson,N.D.(2006).Model system to evaluate the effect of *amp D* mutation on AmpC-mediated beta lactam resistance. Antimicrob.Agents Chemother.50:2030-2037.

Sekowska , A.; Gospodarek , F.and Kaminska, D. (2011) .Antimicrobial susceptibility and genetic similartiy of ESBL – positive *Klebsiella pneumoniae* strains . Arch. Med .Sci. 8 (6) : 993-997.

Shahid, M.; Umesh; Sobia, F.; Singh, A.; Khan, H. M.; Abida Malik and Shukla, I. (2012): Molecular epidemiology of carbapenem-resistant *Enterobacteriaceae* from a North Indian tertiary hospital . N. Z. J. Med. Lab. Sci. 66: 5-7.

Shakil,S.;Azhar,E.I.;Tabrez,S.;Kamal,M.A.;Jabir,N.R.;Abuzenadha,A.M ;Damanhouri,G.A. and Alam,Q.(2011). New-Delhi metallo beta - lactamase (NDM-1):an update.J.Chemother.23:35-38.

Shanmuganathan,C. ; Ananthakrishnan,A ; Jayakeerthi ,S,R ; Kanungo,R.; Kumar,A .; Bhattacharya,S.and Badrinath,S. (2004). Learning from an outbreak:ESBL-the essential points.Indian J.Med.Microbiol.22(4) :255-257.

Sharma,U. K.; Guleria, R.; Mehta,U.; Sood, N. and Singh, S. N. (2010). NDM-1 resistance: Fleming's predictions become true. Inter. J. Appli. Bid. Pharma. Techno. 1: 1244-1251.

Sianou,E.; Kirsto,I.; Pedridis,M.; Apostolidis,K.; Meletis,G.; Miyakis,K. and Sofianou,D. (2011).A cautionary case of microbial solidarity: concurrent isolation of VIM-1-producing *Klebsiella pneumoniae* ,*Escherchia coli* and *Enterobacter cloacae* from an infected wound .J.Antimicrob.Chemother.7:1-2.

Sikarwar, A. S. and Batra, H. V. (2011). Identification of *Klebsiella pneumoniae* by capsular polysaccharide polyclonal antibodies. Inter. J. Chem. Engineer. Applic. 2(2): 130-134.

Singh , R.E; Veena , M.; Raghukumar , K.G.; Vishwanath , G.; Rao, P.N. and Murlimanju , B.V. (2011). ESBL production :resistance pattern in *Escherichia coli* and *Klebsiella pneumoniae* , a study by DDST method .Inter .J.Appl .Biol .Pham. Technol. 2 :415 - 422.

Singhal,S. ; Mathur, T .; Khan,S. ; Upadhyay,D.J .; Chugh,S. ; Gaind,R.and Rattan,A.(2005). Evaluation of methods for AmpC beta-lactamase in Gram- negative clinical isolates from tertiary care hospitals .Indian J.Med.Microbiol.23:120-124.

Siu, L.K.; Lu, P.L.; Chen, J.-Y.; Lin, F. M. and Chang, S.-C. (2003). High level expression of AmpC β-lactamase due to insertion of nucleotides between -10 and -35 promoter sequences in *Escherichia coli* clinical isolates: cases not responsive to extended-spectrum cephalosporin treatment. Antimicrob. Agents Chemother. 47: 2138–2144.

Smet, A.; Martel, A.; Persoons, D.; Dewulf, J.; Heyndrickx, M.; Catry, B.; Herman, L.; Haesebrouck, F. and Butaye, P. (2008). Diversity of extended-spectrum β-lactamases and class C β-lactamases among cloacal *Escherichia. coli* isolates in Belgian broiler farms. Antimicrob. Agents Chemother. 52: 1238-1243.

Snitkin,E.S .;Zelazng, A.M.; Thomas,P.J. ;Stock,F.; NISC comperative sequencing program;.Henderson,D.K.;Palmore,T.N.and Segre,J. A.(2012).Traking a hospital outbreak of carbapenem –resistant *Klebsiella pneumoniae* with whole-genome sequencing .Sci.Transl.Med.4:148ra-116.

Soares, G. M. S.; Figueiredo, L. C.; Faveri, M.; Cortelli, S. C.; Duarte, P. M. and Feres, M. (2012). Mechanisms of action of systemic antibiotics used in periodontal treatment and mechamisms of bacterial resistane to these drugs. J. Appl. Oral. Sci. 20(3): 1-29.

Sobia,F.; Shahid,M.; Singh,A; Khan,H.M.; Shukla,I. and Milk,A.(2011).Occurence of bla_{AmpC} in cefoxitin-resistant *Escherichia coli* and *Klebsiella pneumoniae* isolates from a North Indian tertiary care hospital.N.Z.J.Med Lab.Sci.65:5-9.

Somkiat, P.; Arinthip, T.; and Bhinyo, P. (2007). Conjugation in *Escherichia coli*: a laboratory exercise biochemistry and molecular biology education.Antimicrob. Agents Chemother. 35:440-445.

Souli ,M.;Galani,I. and Giamarellon,H.(2008).Emerging of extensively drug –resistant and pandrug –resistant Gram-negative bacilli in Europe.Eurosurvill.13:1-11.

Souli, M.; Galani, I.; Antoniadou, A.; Papadomiclelakis, E.; Poulakou, G.; Panagea, T.; Vourli, S.; Zerva, L.; Armaganidis , A. Kanellakopoulous K. and Giamarellou, H. (2010). An outbreak of infection due to β - Lactamase *Klebsiella pneumoniae* carbapenemase 2- producing *K. pneumoniae* in a Greek University hospital molecular characterization, epidemiology, and outcomes . Clin. Infect. Dis. 50: 364-373.

Srinivasan, V. B.; Rajamohan, G., Panchdi, P.; Stevenson, K.; Tadesse, D.; Patchanee, P.; Marcon, M. and Gebreyes, W. A. (2009). Genetic relatedness and molecular characterization of multidrug resistant *Acinetobacter baumannii* isolated in central Ohio, USA. Ann. Clin. Microbiol. Antimicrobiol. 8: 21-31.

Stalder, T.; Barraud, O.; Casellas, M.; Dagot, C. and Ploy, M-C. (2012). Integron involvement in environmental spread of antibiotic resistance. Front. Microbiol. 3:1-4.

Stratton,C. W.(2000).Mechanisms of bacterial resistance to antimicrobial agents .Leb.Med.J.48:186-198.

Subha, A.; Devi, R. and Ananthan, S. (2003). AmpC β - lactamase producing multidrug resistant strains of *Klebsiella* spp. and *Escherichia coli* isolated from children under five in Chennai. Indian J. Med. Res. 117:13-18.

Sung, J. Y.; Kwon, K. C.; Park, J. W.; Kim, Y.S. ; Kim, J. M.; Shin, K. S.; Kim, J, W.; Ko, C. S.; Shin, S. Y.; Song, J. H. and Koo, S. H. (2008). Dissemination of IMP-1 and OXA type β - lactamase in carbapenem- resistant *Acinetobacter baumannii*. Korean J. Lab. Med. 28: 16-23.

Syed, T. S and Braverman ,P.K. (2004). Vaginitis in a adolesscents .Adolesc .Med. Clin. 15 (2) : 235 -251.

Tailor, F. (2011). Characterization of *Escherichia coli* and *Klebsiella pneumoniae* with resistance or reduced susceptibility to carbapenems isolated from Canadian hospitals from 2007- 2010. M. Sc. Thesis. University of Manitoba.

Tamma,P.;Savard,P.;Pal,T.;Sonnevend,A.;Perl,T.M.and Milstone ,A.M. (2012).An outbreak of extended spectrum β-lactamase-producing *Klebsiella pneumoniae* in a neonatal intensive care unit .Infect.Hosp.Epidemiol.33(6):631-634.

Tang, L. M.: Chen, S. T.: Hsu, W. C. and Chen, C. M. (1997). *Klebsiella* meningitis in Taiwan: an overview. Epidemiol. Infect. 119: 135-142.

Tato, M. Coque, T. M.; Ruiz- Garbajosa, P.; Pintado, V.; Coba, J.; Sader, H. S.; Jones, R. N.; Baquero, F. and Cantion, R. (2007). Complex clonal and plasmid epidemiology in the first outbreak of *Enterobacteriaceae* infection involving VIM-1 metallo- β - lactamase in spain: toward endemicity? Clin . Infect. Dis. 45: 1171-1178.

Teo,J.;Cai,Y.;Tang,S.;Lee, W.;Tan,T.Y.;Tan, T.T and Kwa, A.L.(2012).Risk factors, molecular epidemiology and outcomes of ertapenem –resistant ,carbapenem –susceptible *Enterobacteriaceae*:a case –case control study .Plos one.7 :e 34254.`

Thirapanmethee, K. (2012). Extended spectrum β - lactamases: critical tools of bacterial resistance. Mahi.Univ.J.Pharm.Sci. 39(1):1-8.

Thomas, L. C. (2007). Genetic methods for rapid detection of medically important nosocomial bacteria. Faculty of Medicine , Department of Medicine, The University of Sydney, Australia.

Thomson, K.S. (2001). Controversies about extended-spectrum and AmpC beta-lactamases. Emerg. Infec. Dis. 7 (2): 333-336.

Thomson, K. S. (2010). Extended- spectrum - β - lactamase, AmpC, and carbapenemase issues. J. Clin. Microbiol. 48(4): 1019-1025.

Thomson, K. S. and Sanders, C.C. (1992). Detection of extended – spectram β - lactamases in members of the family *Enterobacteriaceae*: comparison of double-disk and three-dimensional tests. Antimicrob. Agents Chemother. 36: 1877-1882.

Tijet , N.; Alexander, D. C.; Richardson, D.; Lastoretska, O.; low, D. E.and Patel ,S. N. (2011). New Delhi metallo- β - lactamase, Ontario, Canada. Emerg. Infect. Dis. 17: 306-307.

Tipper, D. J. and Strominger, J. L. (1965). Mechanism of action of penicillins: a proposal based on their structural similarity to acyl-D-alanyl-D-alanine. Proc. Natl. Acad. Sci .USA 54: 1133-1141.

Tokatlidou, D.; Tsivitanidou, M.; Pournaras , S.; Ikonomidis, A.; Tsakris, A. and Sofianou, D. (2008). Outbreak caused by a multidrug-resistant *Klebsiella pneumoniae* clone carrying *bla*VIM-12 in a university hospital.J. Clin. Microbiol. 46 (3): 1005-1008.

Toleman, M. A.; Simm, A. M.; Murphy, T. A.; Gales, A. C.; Biedenbach, D. J.; Jones, R. N. and Walsh, T. R. (2002). Molecular characterization of SPM-1, a novel metallo-β -lactamase isolated in Latin America: report from the SENTRY antimicrobial surveillance programme. J. Antimicrob. Chemother. 50: 673-679.

Toroglu, S. and Keskin, D. (2011). Antimicrobial resistance and sensitivity among isolates of *Klebsiella pneumoniae* from hospital patients in Turkey. Int. J. Agric. Biol. 13(6):941-946.

Tran, J. H.; Jacoby, G. A. and Hooper, D. C. (2005). Interaction of the plasmid encoded quinolone resistance protein *Qnr* with *Escherichia coli* DNA gyrase. Antimicrob.Agents Chemother. 49: 118-125.

Tsakris, A.; Pournaras, S.; Woodford, N.; Palepou, M. F. I.; Babini, G. S.; Douboyas, J. and Livermore, D. M. (2000). Outbreak of infections caused by *Pseudomonas aeruginosa* producing VIM-1 carbapenemase in Greece. J. Clin. Microbiol. 38: 1290-1292.

Turton , J. F.; Kaufmann, M. E.; Glover, J.; Coelho, J. M.; Warner , M.; Pike, R. and Pitt, T. L. (2005). Detection and typing of integrons in epidemic strains of *Acinetobacter baumannii* found in the United Kingdom. J. Clin. Microbiol. 43:3074-3082.

Tzouvelekis, L. S. and Bonomo, R. A. (1999). SHV-type β-lactamases. Curr. Pharm .Des .5: 847-864.

Umadevi, S.; Kandhakumari, G.; Joseph, N. M.; Kumar , S.; Easow, J. M.; Stephen, S. and Singh, U. K. (2011). Prevalence and antimicrobial susceptibility pattern of ESBL producing Gram-negative bacilli .J. Clin. Diag. Res. 5(2) : 236- 239.

Upadhyay, S.; Sen, M.R.; and Bhattacharjee,A.(2010). Presence of different β-lactamase classes among clinical isolates of *Pseudomonas aeruginosa* expressing AmpC β-lactamase enzyme. J. Infect.Dev. Ctries. 4:239-242.

Vaidya, V. K. (2011). Horizontal transfer of antimicrobial resistance by extended- spectrum β - lactamase- producing *Enterobacteriaceae*. J. Lab. Phys. 3: 37-42.

Van Bembeke, F. ; Balzi, E. and Tulkeas , P.M.(2000). Antibiotic efflux pumps.Biochem.Pharmacol.60 (4):456-470.

van- Hoek, A. H. A.; Mevius, D.; Guerra, B.; Mullany, P.; Roberts, A. P. and Aarts, H. J. M. (2011). Acquired antibiotic resistance genes: an overview. Front. Microbiol. 2: 1-27.

Vercauteren,E.   ;Descheemaecker,P.   ;Ieven,M.;   Sanders,C.C.   and Goossens,H.(1997).Comparison   of   screening   methods   for detection of extended spectrum β -lactamases and their prevalence among blood isolates of *Escherichia coli* and *Klebsiella* spp.in a Belgian teaching hospital.J.Clin.Micrbiol.35:2191-2197.

Verwaest, C.; Belgian Multicenter Study Group. (2000). Meropenem versus imipenem / cilastatin as empirical monotherapy for serious bacterial infections in the intensive care unit. Clin .Microbiol. Infect. 6: 294-302.

Villegas, M. V.; Lolans, K.; Correa, A.; Kattan, J. N.; Lopez, J. A. and Quinn, J. P. (2007) . First identification of *Pseudomonas aeruginosa* isolates producing a KPC- type carbapenem – hydrolyzing beta- lactamase. Antimicrob. Agents Chemother. 51(4): 1553-1555.

Vourli, S.; Giakkoupi, P.and Miriagou, V. (2004). Novel GES/IBC extended-spectrum β-lactamase variants with carbapenemase activity in clinical enterobacteria. FEMS Microbiol Lett. 234: 209-213.

Wachino, J.; Doi, Y.and Yamane, K. (2004). Nosocomial spread of ceftazidime-resistant *Klebsiella pneumoniae* strains producing a novel class A b-lactamase, GES-3, in a neonatal intensive care unit in Japan. Antimicrob. Agents Chemother. 48: 1960–1967.

Walsh, T.R. (2008) .Clinically significant carbapenemases : an update Curr. Opin .Infect. Dis.21 (4) :367-371.

Walsh, T. R.; Toleman, M. A; Poirel , L. and Nordmann, P. (2005). Metallo- β - lactamases: the quiet before the storm? Clin. Microbiol. Rev. 18(2): 306-325.

Walther- Rasmussen, J. and Hoiby, N. (2006). OXA-type carbapenemases. J. Antimicrob. Chemother. 57: 373-383.

Wang, X. D., Cai, J. C.; Zhou, H. W.; Zhang, R. and Chen, G- X. (2009). Reduced susceptibility to carbapenems in *Klebsiella pneumoniae* clinical isolates associated with plamid- mediated β - lactamase production and Ompk36 porin deficiency. J. Med. Microbiol. 58: 1196-1202.

Watanabe,M.; Loybe,S.; Inoue,M.and Mistuhashi,S.(1991).Transferable imipenem resistance in *Pseudomonas aeurginosa* .Antimicrob. Agents Chemother.44:1448-1452.

Wei- feng, S.; Jun, Z. and Jian- ping, Q. (2009). Transconjugation and genotyping of the plasmid – mediated AmpC β - lactamase and extended- spectrum β - lactamase gene in *Klebsiella pneumoniae*. Chin. Med. J. 122(9): 1092-1096.

Wiener-Well,Y.;Rudemsky, B.; Yinnon, A.M.Kopuit, P.; Schlesinger, Y.;Broide,E. and Raveh,D. (2010).carriage rate of carbapenem- resistant *Klebsiella pneumoniae* in hospitalised patients during a national outbreak.J.Hosp. Infect.74:344-349.

Wilke, M.S.; Lovering, A.L. and Strynadka, N.C.J. (2005). β-Lactam antibiotic resistance: a current structural perspective. Curr .Opin. Microbiol. 8: 525-533.

Williams , P., and Tomas, J. M. .(1990). The pathogenicity of *Klebsiella pneumoniae*. Rev. Med. Microbiol .1:196-204.

Wing , D.A.; Hendershott, C.M.; Debuque, L.; Millar, L.K. (1999). Outpatient treatment of acute pyelonephritis in pregnancy after 24 weeks. Obstet .Gynecol. 94: 8-683.

Woodford, N.; Tierno, P. M. Young, K.; Tysall, L.; Palepou, M- F., I.; Ward, E.; Painter, R. E.; Suber, D. F.; Shung, D.; Silver, L. L.; Inglima, K.; Kornblum, J. and Livermore, D. M. (2004). Outbreak of *Klebsiella pneumoniae* producing a new carbapenem- hydrolyzing class A β - lactamase , KPC-3 , in a New York medical center. Antimicrob. Agents Chemother. 48 (12): 4793-4799.

World Health Organization (WHO).(1995). WHO Global Strategy for Containment of Antimicrobial Resistance. Geneva, Switzerland.

Wu, J. H.; Wang, L- R.; Liu, Y- F.; Chen, H-M. and Yan , J-J. (2011). Prevalence and characteristics of ertapenem- resistant *Klebsiella pneumoniae* isolates in a Taiwanese university hospital. Microbiol. Drug Resist. 17(2): 259-266.

Yanagawa, T.; Nakamura, H.; Takei, I.; Maruyama, H.; Kataoka, K.; Saruta, T. and Kobayashi, Y. (1989). *Klebsiella pneumoniae* meningitis associated with liver abscess: a case report. Jpn. J. Antibiot. 42: 2135-2140.

Yeoh, K. G.; Yap, I.; Wong , S. T.; Wee, A.; Guan , R. and Kang, J. Y. (1997). Tropical liver abscess. Postgrad. Med. J. 73. 89-92.

Yigit, H.; Queenan, A. M.; Anderson, G. J.; Domenech-Sanchez, A.; Biddle, J .W.; Steward, C. D.; Alberti,S.;Bush,K. and Tenover, F. C. (2001). Novel carbapenem-hydrolyzing β-lactamase, 228 KPC-1, from a carbapenem-resistant strain of *Klebsiella pneumoniae*. Antimicrob. Agents .Chemother. 45: 1151-1161.

Yong,D.Toleman,M.A.;Giske,G.C.;Cho,H. S.;Sundman ,K.;Lee,K. and Walsh, T.R. (2009).Characerization of a new metallo-beta-lactamase gene *bla*NDM and a novel erythromycin esterase gene carried on a unque genetic structure in *Klebsiella pneumoniae* sequence type 14 from India.Antimicrob .Agents Chemother. 53(12):5046-5054.

Younes, A.; Hamouda, A.; Dave, J. and Amyes, S. G. B. (2011). Prevalence of transferable *bla*CTX-M-15 from hospital- and community- acquired *Klebsiella pneumoniae* isolates in Scotland. J. Antimicrob. Chemother. 66: 313-318.

Zakaria, E.A. (2005). Increasing ciprofloxacin resistance among prevalent urinary tract bacterial isolates in Gaza Strip, Palestine. J.Bio.Biotech. 3: 238-241.

Printed in Great Britain
by Amazon

20797650R00129